Philosophy of Science

Hℜ

Philosophy of Science

Liv Egholm

Philosophy of Science

Perspectives on Organisations and Society

Hans Reitzels Forlag

Philosophy of Science

Perspectives on Organisations and Society

1st edition, 1st print run

© Liv Egholm and Hans Reitzels Forlag, 2014

Publishing editor: Martin Laurberg

Copy editor: Kamma Lund Jensen

Translator: Tam McTurk

Consultant: Grace Fairley

Cover design: Harvey Macaulay/Imperiet

Layout and typesetting: LYMI DTP-Service

Type set in: Swift

Printing: Livonia Print

Printed in Latvia 2014

ISBN: 978-87-412-5657-3

hansreitzel.dk

Contents

Preface

This book is intended as an introduction to the philosophy of science in a wide sense. It is particularly aimed at bachelor and master's students at various levels, who need an overview of the different ways of thinking about, understanding and explaining their target field. I consider the philosophy of science to be a general framework that focuses our gaze upon the factors that we believe are essential to describe, understand and explain relationships in the world. It therefore has profound implications for how we explore and study the world.

In this book, I have emphasised the historical and diachronic conditions in the development of philosophy of science perspectives. The different perspectives have always arisen in relation to – and quite often as a response to or reformulation of – previously formulated positions. We can only understand what a perspective entails when we understand what it talks about and opposes. The book therefore has a chronological structure, and emphasises the different perspectives' relationships to each other. As such, in the final chapters, topical questions are identified and raised, and attempts are made to answer them in relation to current research perspectives.

I also hope that the book will help to underline the fact that science and methodology belong together, but that different methods can be employed across the dividing lines in the philosophy of science. Both the methods used to collect materials and the choice of materials were selected on the basis of how the material will be used, what we want to know about (which theoretical perspective focuses our gaze?) and the specific empirical data with which we are working (how sensitive is the topic to empirical study?). It is not the case that particular methods (e.g. collecting through observation or interview) or materials (texts, interviews) belong per se to a specific philosophy of science perspective.

I would like to thank all the colleagues alongside whom I have had the pleasure of teaching in the philosophy of science and methodology, as well as students at the University of Southern Denmark, Aarhus University and Copenhagen Business School, where I have taught philosophy of science topics over the years. Their questions have helped me to refine and clarify what I consider to be significant in the description of the philosophy of science and related topics.

Special thanks to the sixth-semester HA students, Psychology, and second-semester BSc students in Business Administration and Sociology, whom I taught in 2013 and 2014. They have had a crucial influence on the book during the writing and editing process, due to their ongoing feedback on the structure, content and readability of the chapters.

I would also like to thank Antje Vetterlein for agreeing to write a case study about the World Bank that could be used to illustrate the different perspectives throughout the book.

Last but not least, I would like to thank Hans Reitzel Publishers, and especially editor Martin Laurberg, for a very good and inspiring collaboration.

Finally, thank you to my family and friends for your interest, patience and understanding along the way.

May 2014
Liv Egholm

CHAPTER 1

The nature, parameters and concepts of science

The purpose and intention of the book

This book is based on two decades of teaching experience in, and giving thought to, the philosophy of science on bachelor programmes at universities, both in relation to specific subjects (history, sociology, linguistics and anthropology) and on interdisciplinary programmes (languages and communication, information management, business economics and psychology, business economics and sociology, market communications, etc.). The teaching, as well as the students' questions, questioning and feedback, has had consequences for the themes and perspectives that I consider relevant to touch upon, and that transcend the subject-specific traditions. In my 20 years of teaching, I have, of course, read a great number of other philosophy of science and methodology books, as well as primary sources. In different ways, they all serve as background to this work.

The purpose of this book is to present philosophy of science approaches for students on interdisciplinary study programmes. The specific disciplines tend to have a historically elaborated approach to the philosophy of science, which is thus specifically based on this history. This book is not for a specific discipline, but is aimed at the wealth of study programmes (in practice, most humanities and social-science programmes) that combine multiple disciplines. Common to them all is that they have a strong social-science element, with a touch of the humanities. When you cannot rely upon the specific academic tradition that historically epitomises your discipline and its perspectives, it is even more relevant to have an overview of the theoretical approaches that transcend the more subject-specific theories and approaches. The overarching field of interest of many of the interdisciplinary programmes is the organisation. This book is particularly aimed at those programmes; the examples used will reflect this.

It is my experience that students consider philosophy of science both abstract and awkward to deal with, compared to their own and other people's work. To mitigate this, I have chosen to use a single case study throughout the book – that of the World Bank – in order to exemplify more specifically how the seven different philosophy of science perspectives can form the basis for formulating research questions, developing research design, collating material and evaluating the validity and usefulness of analyses. It is my hope that the book will serve as a launching pad for students to reflect critically on how they and others generate knowledge.

To varying degrees, all university study programmes incorporate aspects of the philosophy of science, and with good reason. Studying the philosophy of science presents students with the overall perspectives that underpin the subjects and areas that interest them, and it is from this interrelationship that specific research questions arise and knowledge is generated. Although often very abstract, the philosophy of science is nonetheless one of the keys to discussing and reflecting on how we, and others, produce knowledge and texts.

The different theoretical perspectives are referred to as meta-theories, i.e. theories about theories, representing overarching considerations of the key questions about the nature of being, knowledge and action, which are implicit in the thoughts (theories) we hold about the world. They transcend scientific disciplines. The same meta-theory can form the basis for economic, sociological and psychological theories and studies, despite differences in subject matter. Meta-theories are abstract and generally difficult to apply to specific problems. However, they play a significant role as prerequisites for highly specific aspects of assignments and projects, such as formulating research questions, determining what material is relevant and how to collate it, deciding which analytical approaches to adopt and deciding how to conduct the necessary evaluation of the results of an analysis.

What is a theory and what is it for?

But what exactly is a theory? What can it do? What is it for? The definitions are many and varied, but there is fairly widespread agreement about certain elements. A theory is a lens through which we observe phenomena we wish to understand and explain – for example, social phenomena. In this sense, theories are not the actual social phenomena we are studying. They are abstractions of those phenomena, and help us establish a wider view and generate meaning in the social world. Theories are always abstractions or generalisations of the specific observations that they seek to explain and systematise. In general, theories are used to describe phenomena, explain relationships between phenomena and describe the mechanisms at play in these relationships. A theory therefore leads us to look at the phenomena in new or different ways, helping us ask new and different questions about the subject. As such, a theory should help us to answer empirical, conceptual and/or practical problems.

These empirical problems may stem from something that seems strange or inexplicable, such as the relationship between an organisation and its members. Why do employees not do what their manager tells them to? Theories are also used to solve conceptual problems. These can be internal problems created by theoretical inconsistency, or external problems where two theories conflict. For example, if you are studying theories about both open and closed organisations, then you need an overall theory capable of accommodating all of the different elements. A theory can also serve to clarify what it means for individuals to be understood purely in terms of their relationships.

Theories are sometimes also used to solve practical problems. As the social psychologist Kurt Lewin put it: "There is nothing as practical as a good theory" (1951: 169). Theories can be applied to questions such as: Why do all organisations in the pharmaceutical industry resemble each other? And is this phenomenon conducive to profitability? Within a large international organisation, how is it possible to build a team that transcends national borders? Here, a theory (assuming it is good) may act as a guide or help provide explanations about things we do not understand.

Different levels of theories

There are many levels of theories. They have different functions and can explain different levels of phenomena. These levels can be roughly split into three: *meta-theories, general theories* and *specific theories*. All of these levels are usually involved when you read or produce an academic text or write a project.

Meta-theories are abstract, general and quite difficult to deploy directly in a study of a particular phenomenon. They are overall worldviews based on particular ideas about the basic nature of reality (ontology) and how we seek to understand it (epistemology). They delineate the basic conceptions behind questions such as: What kinds of phenomena are interesting? Why are they relevant? What is it we want to know about? Which research questions are interesting and relevant? How can we collect and analyse relevant material, and answer the question? As a result, meta-theories are best identified through specific themes, questions and problems.

General theories explain and describe more specific themes. These may be theories of identity, communication, culture, functionality, etc. They are primarily used as models to explain or identify readily overlooked elements of the phenomena being studied.

Specific theories explain or describe a small group of more specific phenomena. They are limited in scope, and linked to specific contexts. They may be theories about how managers identify themselves in companies, communication in hierarchical organisations, organisational culture in merged companies or the functionality of public websites. Although much more specific, these theories help formulate new questions that can be posed in contexts not previously described or analysed on the basis of the specific theory.

The phenomena and elements we study consist of many different interrelationships. Theories ensure that certain relationships are more prominent than others. Our theoretical perspective, be it general or more specific, zooms in on the relationships in which we are particularly interested, e.g. identities in organisations, even though organisations, of course, contain many other aspects and relationships on which we could just as easily have focused.

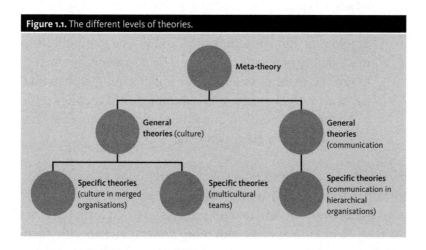

Figure 1.1. The different levels of theories.

How are the various levels of theories used?

General and specific theories can be applied differently depending on the purpose of the research or project. Generally speaking, there are three ways of working with theories in research and projects:

- The first method uses theories to explain and understand the phenomena in which we are interested – usually, these are general theories that concern parts of the phenomena being studied. The theories are used to raise new questions that illustrate, group and perhaps compare the observed phenomena in a new and different way that creates new understanding of the phenomena.
- The second method tests existing theories on relevant analytical material. The idea is to identify strengths and weaknesses in existing theories, and perhaps identify how the theory might be developed or what should be taken into account if the theories are to be used to explain the world. Specific theories are usually central to this approach.
- The third method is theory development by means of research and project writing. Meta-theories or general theories are typically the starting point for this process. The phenomena being studied are described on the basis of the ontological and epistemological conditions that underlie a meta-theory or a general theory. The purpose is not merely to use the theories to explain or understand the phenomena, but to develop new understandings and theories (typically specific ones, but sometimes general theories) that can be applied in other contexts.

Meta-theories are perspectives in the philosophy of science, and form the focus of this book. Although meta-theories are rarely cited as the basis for articles and projects, they do form the basis for the use of both specific and more general and specific theories. It is therefore essential to understand the implications of the various perspectives for research and project-writing.

Generating knowledge and applying methodology

As mentioned previously, the philosophy of science perspectives inherent in meta-theories help to identify relevant and interesting subjects, problems and research questions. However, it is one thing to ask questions; how you find the answers is another matter, and depends on which method you use to study the world. In order to understand or explain phenomena and the relationship between them, you must recognise that the way in which you see the world is determined by one or more implicit theoretical perspective(s). Here too, the perspectives play a key role.

For example, to pose questions about a group of people's attitudes to change in the workplace, you have to think about how you will elicit the information from them. Should you set up interviews or read their own personal accounts (if these exist)? Or should you observe how they behave in everyday situations? What questions/documents/situations would it be relevant to pose/read/observe in order to answer these questions? And what problems might that involve? With whom, about what and how should you conduct interviews/read documents or make observations? How can you say something credible about the problem you wish to illustrate? And how do you ensure that you do not misunderstand the group's experiences, distort them or only articulate the views of some of them? In other words, the relationship between your worldview (what you consider important/relevant) and the real-life situation being studied is crucial. This relationship helps to determine what questions you ask, what design you adopt for your research, what material you collate and how, and how you subsequently analyse the material and evaluate the impact of the analysis (how certain are your conclusions?) and the range of your propositions (how general are your conclusions?).

Quite a few different methodological approaches and techniques can be deployed to collate and analyse material. They often emerge from particular theoretical perspectives, but that does not necessarily mean that they are only applicable in this context, or that all studies making use of these methods and techniques are necessarily based on the same perspective. In order to understand their connection to the theoretical perspectives, it is important to keep in mind how, and for what purpose, the various techniques and methods are deployed.

As such, the more abstract philosophical perspectives always play an active role in addressing and reaching conclusions about scientific questions at all levels. Awareness of these perspectives helps improve your ability to evaluate your own knowledge production and that of others.

What is science?

Science has not always been a special, separate domain with its own particular logic. As late as the 18th century Enlightenment, art, faith and science co-existed happily. Post-Enlightenment, the natural sciences dominated the definition of science. As a result, science and scientific tools for evaluation have to a great extent historically been defined on the basis of the natural-science ideal of making general statements about general laws.

Science is embedded in both tangible and intangible institutions. Tangible institutions include, for example, universities, disciplines and the role of science in society. We consult experts on all sorts of issues and use their opinions to justify important social decisions. Our starting point is therefore that scientific knowledge is more credible than other forms of knowledge. This also means that researchers generally enjoy a particularly privileged position in our society.

The intangible institutions set norms and rules for how researchers behave. In overarching terms, science is about studying phenomena carefully, honestly and systematically. However, the degree of specificity in the norms and rules for achieving this, and for drawing a demarcation line between scientific knowledge and non-scientific knowledge, depends on which perspective we adopt as our starting point.

These norms are again defined on the basis of how we think about science. There is no single science or way to understand science; there are several, dependent on which structuring principles are applied. Three approaches are introduced/discussed below.

Unity of science

Unity of science is a term used to describe the perception that all forms of science are ultimately rooted in the same basic principles. They are inspired by the natural-science tradition, so take objective empirical studies as their starting point to generate and define theories and propositions. The unity of science can be based on both ontological reduction and methodological reduction.

- Ontological reduction implies that, overall, all objects share the same form of being – they are essentially the same, and therefore can be studied in the same way. The assumption is that there is basically no difference between, say, ideas, gravity, soil conditions and social movements.
- Methodological reduction assumes that even if all subjects are not ontologically identical (if, for example, you assume that there are substantial differences between ideas, gravity, soil conditions and social movements), they must still be studied in the same way in order to generate scientific knowledge. This moves the focus of science from the subject being studied to the way in which it is studied.

The German philosopher and historian of ideas Wilhelm Dilthey (1833–1911) criticised both ontological and methodological reduction and the resulting scientific demarcation. Based on the human sciences (Geisteswissenschaften), Dilthey claims in his Einleitung in die Geisteswissenschaften (Introduction to the Human Sciences, 1883) that the natural sciences and the human sciences study radically different areas, which cannot in any way be said to have the same ontology and therefore should not be studied in the same way.

He points out that there is a fundamental difference between the epistemological purpose of the natural sciences – namely, explaining and predicting causal regularities – and that of the humanist tradition, which aims to understand the meanings and importance of human products (thus rejecting ontological reduction). For Dilthey, it is absolutely central that the various subjects cannot be studied in the same way, but require methods tailored to the different areas (thus rejecting methodological reduction). The non-human and the human should therefore be studied differently. Natural scientists are supposed to *explain* their subject, while humanities researchers are supposed to *understand* theirs.

The traditional division of the sciences into human vs. natural sciences

Dilthey's division of science into two different approaches, each with its own subjects, methodology and material, made a significant impression on the scientific understanding and on the general understanding of science. Typically, the division was between the "soft" human sciences and the "hard" natural sciences. The terms hard and soft denoted both the type of material traditionally analysed, and the way in which it was processed. Typically the "soft" subjects used words or other material, which were qualitatively and interpretively processed, whereas the "hard" subjects used numbers or formulas that could be counted quantitatively and mathematically. This division of the sciences and their methodologies was based on a whole range of assumptions, as shown in Figure 1.2 (Smith 2005: 110).

The natural sciences	Humanities
• Ethics (seeing and describing a phenomenon from the outside)	• Emic (the phenomenon is seen and described from an insider perspective)
• Observer perspective	• Participant perspective
• Tangible objects and unintended phenomena	• Intended, meaningful actions, texts and artefacts
• Careful observation, accurate quantification and calculation	• Careful studies, close reading and rigorous reflection
• Descriptive, value-neutral	• Normative, critical
• Collating materials, experiments, establishing and testing hypotheses	• Archive research, text and concept analysis, interpretation
• Seeking knowledge, prediction, checking	• Seeking understanding, insight, evaluations
• Using models, causal explanations, general theories	• Using analyses, comments, criticisms, reconstructions
• Indicating rules, facts, laws	• Indicating relationships, values, possibilities
• Emphasising simplicity, clarity, quantification	• Highlights richness, depth
• Replicable by others	• Recognised by other researchers

Figure 1.2. The natural sciences and the humanities.

The consequences of this dualistic division include that the scientific nature of the humanities – and later on, also the social sciences – was considered problematic. As a result, science had to be enhanced, and this was typically achieved by methodological reduction, in other words by becoming more like the natural sciences.

Tribal warfare

This dualism in the academic landscape has led to significant differences being formalised, both between and within the individual disciplines. This is particularly true in disciplines typically thought of as social sciences, such as sociology, psychology and anthropology, but also those in the humanities, such as literature, linguistics, history, etc.

The British natural scientist C.P. Snow, in *The Two Cultures in Science* (1959), describes the differences in scientific culture as extending even to the appearance and behaviour of the academics involved – and traces of this division are still visible. Up until the 1970s, the boundaries were generally considered to be *between* the different subject areas, for example, between literature and economics, but it is symptomatic that what Barbera Herrnstein Smith, in *Scandalous Knowledge*, calls "tribal warfare" (Smith 2005: 108ff) is now typically waged *within* individual disciplines. For example, sociologists and psychologists are more different from each other than they are from practitioners in other areas based on the same type of science, general subjects and epistemological focus. In other words, quantitative sociologists are more similar to quantitative psychologists than qualitative sociologists, even though the individual discipline will have points of focus that are within the same general subject area. An analysis of causality in psychology may focus on how the individual reacts to external stimuli as rewards; a sociological analysis may look at the relationship between women in work and institutional frameworks; and an economic analysis might look at the causal relationships between production and mergers. It is crucial to understand that within each discipline there can, of course, be radically different perspectives on what it is relevant to study, and which materials and methods will ensure valid and useful answers to the question posed. Although the specific disciplines studied are concerned with specific topics, their overall interest in, for example, causal connections and predicting phenomena may be consistent with other disciplines concerned with other topics.

Epistemological focus on the demarcation line between sciences

The German philosopher and sociologist Jürgen Habermas (1929–) discusses Dilthey's dualistic division into two types of science in his book *Erkenntniss und Interesse (Knowledge and Human Interest,* 1968). Habermas' starting point is that it is not substantial (ontological) and subject-specific differences that characterise the different types of science, but rather their position in modern society. In order to identify the different sciences, he therefore generalised and grouped them around their knowledge-constitutive interests and the subjects associated with them. He was then able to outline three classical types: natural, human and social sciences (Habermas 1968a: 312–317).

- Habermas defines the subject of the natural sciences as nature or the part of reality that is not man-made. The epistemological focus here is technical, as the aim is to establish general *(nomothetic)* laws about causal connections through an *empirical-analytic* approach, and make it possible to predict future events.
- In contrast, Habermas identifies the subject area of the humanities as mankind, its culture and knowledge – areas typically mediated through language. The epistemological interest is defined as *hermeneutical* (interpretative). The interpretation of texts, understood in the broad sense (including images, sayings, social conventions, etc.), is considered crucial to this idea of science, which is oriented toward the individual's own understanding rather than objective conditions. On top of this, it is concerned with what is peculiarly human, which is to be understood in its uniqueness *(ideographic)* and therefore can provide a guide to actions, which also makes the epistemological focus *practical*.
- Habermas characterises the subject area for the social sciences by its interest in human collectives and institutional conditions. The focus is therefore on the more general *(nomothetic)*. In Habermas' interpretation, the social sciences' epistemological focus revolves around general laws – the aim being not to predict, but to be *critical and emancipatory*. The social sciences are thus supposed to identify which laws and structures are real and immutable, and which are merely ideologically based and changeable. The social sciences should therefore have a political and *emancipatory* effect on society.

It is important to stress that Habermas' division into the three sciences does *not* represent a division into subject areas or disciplines, such that sociology is exclusively characterised as a social science, history as one of the humanities. Rather, Habermas' division relates specifically to the knowledge-constitutive interests. In sociology or psychology, for example, we find natural-science, human-science and social-science knowledge-constitutive interests.

Philosophy of science perspectives

The different perspectives in the philosophy of science can be divided in many different ways. For example, beneath Habermas' three main directions (the positivist, the interpretive, and the critical), there are myriad sub-perspectives. They can also be divided into so many different perspectives that the single individual perspectives no longer have any real meaning as overarching concepts.

Although the division into three main directions makes sense in terms of enabling a general identification of the currents in question, it can also group together perspectives whose differences are greater than their similarities. The perspectives can be identified as different -isms. However, a fully-fledged -ism cannot be described without shaving off the odd edge here and there, in order to draw a relatively clear picture of that distinct -ism's subject field and view of knowledge. This means that the details and the small differences between the various thinkers who represent the individual -isms may be smoothed away. Thus, the description of the individual perspectives in the philosophy of science serves to balance out the differences and similarities that we consider essential. Consequently, different traditions within philosophy of science literature stress different aspects of the perspectives.

Key concepts in the philosophy of science

Much like the definition of science, there are also several key concepts in the philosophy of science. This book focuses on four of them: *ontology* (What is the fundamental nature of phenomena in the world?); *epistemology* (How can I know them?); *anthropology* (What view of humankind is inherent in the perspective?); and *value freedom* (Can knowledge be objective in the perspective?). These concepts are the key to identifying the different philosophy of science perspectives' basic indicators. For each concept, I will provide a brief account of the characteristics in relation to which we can assess the different philosophy of science perspectives. These are general concepts with a range of different elements. They can be used as analytical tools to identify and compare the different philosophy of science approaches that we meet in, and apply to, scientific work.

Ontology

Realism and constructivism

Ontology is one of the key concepts in philosophy of science. It means the science of being, and stems from the Greek *ontos* (being) and *logos* (science or "study of"). The question of which ontology underpins a specific perspective is one that addresses the basic assumptions about the nature of the (social) world upon which the individual perspective is based. The philosophy of science focuses on the phenomena that are regarded as key to understanding and explaining the world and its changes. Therefore, the individual perspective's ontology will be clearly linked to its overall subject area and epistemological theory.

One of the main ontological distinctions is between realism and constructivism. Where realism assumes that objects, relationships and characteristics exist in the world independently of our understanding of them, the constructivist approach posits that we can only study our own understanding of objects, relationships and characteristics. Therefore, in a realistic ontology, social phenomena have an existence in and of themselves, as opposed to constructivist ontology, in which they can only be studied in terms of the meaning the observer attributes to them. Where *ontological realism* seeks to understand and explain that which exists in the world independently of how academics understand it, *ontological constructivism* focuses on how these factors are created and perceived. This applies to such diverse phenomena as causal relationships, actions, ideas or mentalities. For an elaboration of the difference between realism and constructivism, see Chapter 7.

Materialism versus idealism

Where realism and constructivism consider the role of human consciousness and cognition in relation to the nature of the world, there is also another relevant division of the ontological perspective – namely, between materialism and idealism. At an elementary level, the division is about the elements that form the basis of our explanatory framework. Which elements are considered dominant – the material elements, which have tangible importance; or the more idea-based elements, which account for phenomena and their development?

In the idealistic perspective, the intellectual dimension – our thoughts, God, freedom, culture or rationality – prevails as that which moves mankind to act as it does. Material conditions emerge from culture, ideas and thoughts. One of the best-known idealists was the Greek philosopher Plato (427–347 BC), for whom ideas have their own objective and independent existence – indeed, they are the essence of reality, whereas the material world is merely its shadow.

The opposite of this is materialism, which claims that ideas and phenomena of consciousness always emerge from the physical and material conditions in the world. This contrast is illustrated by the difference between the sociology of Karl Marx (1818–1883) and Max Weber (1864–1920). Marx argues in *Das Kapital* I–III (1867–1894) that the material is the motivating force in society, and therefore that material conditions should be the focal point of academic analysis. Marx divides society into what he calls *base* and *superstructure*. Base refers to the economic and material structure (the worker, the means of production and the owners) that determines the fundamental frameworks for society and characterises the content of the superstructure (culture, politics and ideology) (for an elaboration of this, see Chapter 6). In his book *Die Protestantische Ethik und der Geist des Kapitalismus (The Protestant Ethic and the Spirit of Capitalism, 1904)*, Weber turned this concept upside down. He claimed that the ideas, semantic ordering and ethics that lie behind human actions establish society's frameworks and thereby determine the possibility of organising the means of production and utilising specific resources.

Universalism, context and process

This ontological distinction between the real and the hypothetical is not the only key question in relation to the ontology. Another significant question is whether we perceive phenomena as universal, contextual or processual.

Can the phenomena we study be considered *universal?* In other words, do they remain the same no matter where and in what time period we study them? For example, is an organisation or a form of management the same phenomenon in Denmark and China? And was it the same phenomenon in 1795 as it is in 2014?

Or do we need to understand phenomena in their temporal, spatial and cultural *context* in order to understand and describe them, so that we consider an organisation in China as something distinct from an organisation in Denmark, or consider an organisation in 1795 to be something different from one in 2014?

We must also consider whether phenomena are static or dynamic – and if they are dynamic, what are the motivating forces behind them? Can phenomena be explained and described as delineated, stable entities, or should phenomena be considered as undergoing continuous rebirth, and therefore as *processual?* Can an organisation be studied as a unit, either universally or contextually, or is the organisation continually recreated and, as such, must be understood and explained as a process?

These and similar questions can be raised about both your own and others' analyses and theories, and will help you to identify the perspective that forms the basis of the analysis or the theory's arguments.

Epistemology

Objectivity, subjectivity and intersubjectivity

Epistemology is another key concept in philosophy of science. It means the study of the theory of knowledge, and stems from the Greek *episteme* (science, knowledge) combined with *logos* (science or "study of"). Epistemology is about the nature of knowledge, how we know something, and how knowledge can and (perhaps) must be produced. The epistemology of each philosophy of science perspective indicates the underlying hypotheses that are used to assess knowledge and to determine, by extension, whether that knowledge is credible and valid. A central epistemological distinction is made between subjectivity and objectivity, i.e. whether it is possible to achieve objective knowledge. This question is closely related to what we mean by truth – not least, how we identify one thing as true and another as untrue (this concept will be discussed in greater detail in Chapter 2). Bluntly, the epistemological objectivity criterion can be described as the question of whether we can acquire knowledge about the world as it is, without that knowledge being distorted by human cognition.

The German philosopher Immanuel Kant (1727–1804) argues in *Kritik der reinen Vernunft* (*Critique of Pure Reason*, 1781) that humankind does not have direct access to knowledge of the world and of things through cognition. In doing so, he underlines the central discussion about the relationship between what the world is and what we as humans are able of knowing of it, a discussion that is still ongoing. Kant calls the world and things in it that are not yet perceived by humans *Das Ding an sich* (the thing itself, often translated as *noumenon*), and calls the world and things in it that are perceived by humans *Das Ding für uns* (the thing for us, often translated as *phenomenon*). Very few would disagree completely with the difference between *Das Ding an sich* and *Das Ding für uns,* and as a consequence most academics consider complete objectivity an ideal.

Objectivity would entail an attempt to ensure a complete absence of observer bias with regard to a phenomenon as well as unambiguous comment on it. To achieve this, the individual who seeks to acquire knowledge would have to be methodically separated from that which he or she wants to learn about. The relationship can be described as follows: a subject (the academic) looks at an object (that which he or she wants to know something about, which can be separated from the academic and considered in isolation). This is known as *methodological objectivity.* In contrast to this is the more subjective approach. As described above, the subjective approach has typically been criticised as unscientific, and as one of the main failings of the humanities. The key point here is that the person who wants to know something, cannot be separated from what she or he wants to know something about. The interpretation of texts, signs, interviews, behaviour, etc., cannot take place without the researcher – the person who wants to know – necessarily being connected with that which he or she wants to know something about. The relationship is that a subject (the interpreting academic) looks at a subject (that which he or she wants to know something about, which cannot be separated from the interpreting academic). Although the subjective approach has been strongly criticised, many researchers point out that it is not only impossible to decouple the observer from that which is observed, but that the actual connection between the person observing and the object observed is a necessary and useful condition of the research. This is called *methodological subjectivity.*

One attempt to avoid the subdivision into either purely objective or purely subjective knowledge is known as intersubjectivity. Intersubjectivity combines individual subjective understanding (that all analyses are based on individual subjective interpretations) with the opportunity to check and validate results and analyses. Intersubjectivity involves multiple competent individuals, using the same analyses and interpretations, being able to arrive at (approximately) the same result. This means, therefore, that the researchers must be careful to present every part of their analyses and procedures, so that others are able to follow in their footsteps and check the results. Validity is not achieved by the result corresponding with reality, but by agreement between different researchers, which raises the analysis from an individual, subjective level to an intersubjective level.

Empiricism versus rationalism

A related question is how we acquire, or are supposed to acquire, knowledge. In this context, there are two key starting points: *empiricism and rationalism*. Empiricism stems from the idea that knowledge is based on sensory experience. Our knowledge is therefore *a posteriori*, i.e. it emerges *after* the encounter with the world. The early Anglo-Saxon philosophers Francis Bacon (1561–1626), John Locke (1632–1704), John Stuart Mill (1806–1873) and Scotland's David Hume (1711–1776) were among the foremost empiricists. They were all interested in how knowledge could be achieved through sensory experience, as well as the limits imposed by the sensory apparatus.

A key criticism of the endeavour to base scientific findings on sensory experience is that it is based on the hypothesis that observations are unaffected by the context in which they are made or understood. This is precisely the criticism raised by the rationalists. The rational position stretches back to the Greek philosopher Plato (428/427–348/347) via the French philosopher and mathematician René Descartes (1596–1650) to Kant (1727–1804). The idea is that all knowledge arises rationally and *a priori*, i.e. before the senses, in the light of the human reason that structures what we know. In *Meditationes de Prima Philosophia (Meditations on First Philosophy*, 1641), Descartes argues that we must start out from the hypothesis that true knowledge cannot emerge from our sensory experiences, the potential distortions caused by which are easy to identify. Descartes sets out to doubt everything that can be doubted, which leads him to scepticism: "Whatever

I have accepted until now as most true has come to me through my senses. But occasionally I have found that they have deceived me, and it is unwise to trust completely those who have deceived us even once." In doing so, he posits doubt *(methodical doubt)* as the ultimate basis for certain knowledge. The famous quote *"Cogito ergo sum"* (I think, therefore I am) denotes that the only thing Descartes did not doubt was that he actually existed.

Epistemological questions

In extending the distinction between rationalism and empiricism, the question arises of what material we must collate and process in order to acquire knowledge about our subject. This raises, among other things, the following questions: Should scientific knowledge describe, explain or understand phenomena, or do all of these at the same time? And is it through interpretation, observations or quantification that we derive the most reliable knowledge of the world? What forms of conclusions are used? Which are considered useful? Should we draw conclusions from individual cases to general rules *(induction)* or from general rules to individual cases *(deduction)*? Or look at the relationship between general rules and individual cases (abduction/retroduction)? (Induction and deduction are discussed in greater detail in Chapter 3, retroduction in Chapter 6 and abduction in Chapter 7.) All of these questions are important for determining which methods you must use to obtain the most reliable/true/best-possible knowledge of the phenomena you study. The choice of method is inextricably linked to the epistemology, because the more abstract question of how you can acquire knowledge about the phenomena you study controls your specific choice of material, method of collation and form of analysis.

Anthropology

What is a human being?

All perspectives in the philosophy of science encompass ideas of what constitutes a human being. This involves, for example, the question of what drives people to act as they do. Do we act through rational choice? Are human beings rational individuals? If so, what does this rationality consist of? Is it universal, or does it stem from cultural and historical context? Will people strive to achieve a maximisation of their material or emotional needs? Are human motivations and behaviour fundamentally driven by material, emotional or social exchanges? Or do personal desires and needs drive the individual? And if they do, can we talk about forms of desire economy (is the desire rational, like our economic behaviour)? Are the forces that govern us conscious or unconscious? These, and many more, questions are implicit in both the specific description of phenomena and more generally in the philosophical perspectives in question. It is not always clear which view of humanity applies in the individual approaches but, by studying descriptions and explanations of, for example, changes, it is possible to detect the view of humanity inherent in both philosophy of science perspectives and concrete analyses.

Who is the actor – the structure or the individual?

The question of what drives people to act as they do in turn leads us to the question of who or what is the actual actor – the structure or the individual? The humanities and the social sciences deal with phenomena that, in one way or another, concern human actions, in relation either to the individual or to multiple individuals in group-like formations. It is therefore essential to look at how the individual's rationality, desire or craving to maximise is linked to the social conditions in which they exist, in order to understand why phenomena occur and why they change. Who or what is the acting unit, the individual or the structure? Some approaches would argue that individuals, their intentions and their efforts to maximise their utility comprise the actor behind the actions and changes. From this perspective, a financial crisis can be seen as a result of greedy individuals trying to optimise their potential and material wealth. Other approaches would argue that individuals do not even exert control over their own desires and rationales, but that these are generated by the social or financial structures in which they live, or the functions of which they are part. In this perspective, the financial crisis could be explained in terms of how economic structures/social dynamics led to potential options, which in turn led to individuals acting as they did. The crisis was therefore created by the system/the social dynamics – not by individuals' actions, desires or need for satisfaction. In this view of human nature, the structure or society has agents, and can therefore be studied on this basis. Several researchers have attempted to construct different combinations of these two extremes, in which structure and individual agency are considered coherent and relational. In these combinations, it is impossible to separate individuals from the social situations in which they exist, but that does not mean that individuals' actions can be explained solely through these (social) structures.

Individuals or collectives

The question of whether the actor is the individual or the structure also influences how we study events and changes, and how we select the unit of study. The view of human nature that considers the individual as an actor understands society as an aggregation of individuals. This means that society and events are studied on the basis of the individual's perspective, and we must look at individuals collectively – or alternatively, as representatives of all individuals' driving forces – in order to study and understand phenomena and changes. An opposing view of humanity assumes that we cannot understand the individual in his or her own right, as the individual is always part of a collective. Therefore, when we talk about and study society, we must base our work on the collective. In this context, it is not possible to consider the individual as representative of him- or herself or of other, similar individuals, but as representative of the collective. Both perspectives are therefore able to study the individual, but for different reasons. For the individual view of human nature, it is about understanding the individual, while the collective view of humanity is about gaining access to the collective elements that the individual represents. In the relational view of humanity, both the collective and the individual are present at the same time.

Value neutrality (axiology)

The fourth philosophy of science concept is *axiology*, i.e. the study of values. In this book, the concept is used to examine the perspective adopted by the various approaches to whether value neutrality is necessary and possible. In general, most scientists think that values should not influence academic practices. Most theoreticians and researchers believe, however, that value neutrality in the research process is actually impossible. Several influential philosophers have argued that it is impossible to avoid a certain value attribution in the research process. Therefore, the discussion is not ultimately about value freedom, but more about how values are a part of, and shape, scientific practice. We can outline three different positions.

The first identifies the need to distinguish between the context in which the questions are formulated, and the context in which hypotheses are critically tested for validity. A central figure in this context is the German theorist Karl Popper (1902–1994) who, in his 1972 work, argues that we cannot eliminate values from research questions. He points out that they typically emerge because our experiences of the world inspire us to raise certain questions. However, we should avoid research bias by adopting methods that allow us to ensure that the study and testing of conclusions and hypotheses are not affected by the ideas that triggered the study in the first place (for more details, see Chapter 3).

The second position is that we cannot distinguish between research questions and the study of them. We must take as a starting point that the value attribution in the questions is also important for the way in which we collate and analyse material in order to answer questions and assess the credibility of our conclusions. Therefore, we cannot say that parts of our research are exempt from value attribution, but we can be aware of the consequences and take them into account when evaluating the credibility of a study.

The third position also recognises that value attribution occurs throughout the research process. Unlike the other two positions, it asserts that a political and value-based starting point is actually necessary in order to achieve the research objective, i.e. to create change, stimulate justice or provide silent groups with a stronger voice. This, of course, has certain consequences, as pointed out by the social constructivist researcher Egon Guba:

> If values do enter into every inquiry, then the question immediately arises as to what values and whose values shall govern. If the findings of studies can vary depending on the values chosen, then the choice of a particular value system tends to empower and enfranchise certain persons while disempowering and disenfranchising others. Inquiry thereby becomes a political act. (Guba 1990: 24)

Structure of the book

How do we differentiate between perspectives in this book – and why?

Seven different perspectives are presented in the book. They are chosen because they play a key role in contemporary debates about what constitutes the nature of science. The seven perspectives are: positivism and critical rationalism (Chapter 3); hermeneutics (Chapter 4); phenomenology (Chapter 5); structuralism (including structuralist Marxism and critical realism) (Chapter 6); social constructivism (Chapter 8); pragmatism (Chapter 9); and actor-network theory (Chapter 10). In addition, the book contains three chapters on key basic concepts, fundamental debates and themes that have consequences for the different perspectives: the nature of science and its characteristics (this chapter); the question of different paradigms and the definition of truth in the sciences (Chapter 2) these two chapters set the framework of the book, defines the concepts used to identify the different philosophy of science perspectives. The divide between realism and constructivism as a central turning point in the development of philosophy of science perspectives (Chapter 7) is discussed in the middle of the book too show how a constructivist approach is developed through radicalism of some of the more realist perspectives.

The seven perspectives will be presented in relation to four key concepts: *ontology* (the basic nature of phenomena): *epistemology* (how can we acquire knowledge of them?): *anthropology* (which view of humanity is inherent in the perspective?): and *value-freedom* (can knowledge be objective according to the perspective in question?). The concepts will be used as parameters to identify and highlight key elements of the different approaches. We can identify the same ontological or epistemological approach in different and yet similar perspectives. At the same time, the concepts parameters will make it difficult to place the different perspectives on one or the other end of the line (e.g. realism and constructivism), placing them somewhere in-between instead.

The historical perspective

It is important to remember that many of the -isms presented in this book are not just a product of their own time, but have also evolved in response to each other – in the same way as the concept of the unity of science, Dilthey's dualism and Habermas's tripartite division all emerged in response to previous perceptions of the nature of science. The various -isms will therefore be presented chronologically, even though they all still play a key role in contemporary discussions. Some of the -isms (e.g. pragmatism) emerged long ago (in the late 19th century), only to retreat into oblivion for a period and then re-emerge with renewed vigour in the 1990s.

Historically, a number of important pioneers and philosophers are considered figureheads for the various-isms. The most important individuals and their ideas are introduced at the beginning of each chapter. After this, each chapter looks at the philosophy of science perspectives according to the four main parameters: ontology, epistemology, anthropology and value-freedom. I have chosen to associate the description of the different perspectives with figureheads, in order to show that individual thinkers can have more nuanced and specific interpretations of the perspective, which may be omitted from the more general picture. This also shows how individual pioneers also effectively serve as identity markers, thereby indicating the philosophy of science perspective(s) that the individual researcher uses as his or her starting point.

The book's ambition is to link the more abstract theoretical insights to tangible consequences for research and analysis. For this reason, the book has a recurring case study, which is described directly after this introductory chapter. Following the more theoretical descriptions of the different directions, the case study will be used to describe in real terms the problems that are typically raised by the perspective described (focusing on the parameters in the concepts of ontology and anthropology), and how – and on the basis of what material – it would typically be possible to respond to these problems (focusing on the parameters in the concepts of epistemology and axiology, and concepts of truth).

The case study looks at the way in which the World Bank (WB) has developed in organisational terms, and its explanation and combating of poverty in the period 1970–2000. It was written by Antje Vetterlein, who has researched various aspects of the World Bank for decades.

A Changing Organisation:
The World Bank's Development Strategy From The 1970's Untill Today

A case study of the World Bank

by

Antje Vetterlein

In the past two decades, the World Bank has intensified its fight against poverty. This is best embodied by the *Poverty Reduction Strategy Papers* (PRSP) initiative, adopted in 1999, which signifies a shift towards a more holistic understanding of development. In fact, the Bank has changed its approach to development several times during its existence. In the 1970s, development was understood as enhancing technology and ensuring that basic human needs were met, while, in the 1980s, the policy focus shifted to structural adjustment lending (SAL), and the political conditions necessary for development policies to work.[1] In the mid-1990s, the focus shifted again, to what is often called a holistic understanding of development. The main difference between these perspectives lies in the relationship between, and the relative significance afforded to, economic growth and poverty reduction, the two main objectives of development. Specifically, if the main focus of development is on growth, economic policies will be given priority *vis-à-vis* social policies. Conversely, if poverty reduction

---

1 SAL also marked a shift from project to programme lending. The Bank's primary activity had been project lending, i.e. financing specific development projects such as the building of a school, or infrastructure projects such as dams or roads. Programme lending, on the other hand, refers to overall loans given to the government of a country in times of liquidity problems. In general, the lending process always starts with the country turning to the World Bank for a loan. In close communication with the government concerned, the country office at the Bank develops the project/programme and negotiates the loan conditions. Final approval for the loan rests with the Board of Executive Directors.

is more important, and is seen as a precondition for economic growth, then social policies will be prioritised.

Social and poverty issues surfaced quite early in the Bank's history, as it was initially set up as an aid organisation. In particular, Latin American countries pushed for the inclusion of "development" in the Bank's mandate before signing the *Articles of Agreement* at its foundation in 1944. However, the Bank spent its first few years administering the Marshall Plan because the focus was on reconstructing Europe after the Second World War, rather than on developing countries. At the time, the Bank was mainly a financial organisation. This changed during the 1970s, when more and more developing countries turned to the Bank for financial assistance because they had been hard hit by the economic turmoil of the 1970s, including two oil crises and worldwide stagflation.

Given that its mandate also referred to "development", the Bank had to act and respond to these new demands. However, there was a lack of expertise in these areas. In this respect, Robert McNamara (president from 1968 until 1981) played a crucial role, and transformed the organisation in many ways. His most prominent achievement was the nearly tenfold increase in International Bank for Reconstruction and Development (IBRD) and International Development Association (IDA) lending (from $1 billion in 1968 to over $12 billion in 1981), equivalent to almost a fourfold increase in total commitments in real terms (Lateef 1995: 22). During McNamara's time at the Bank, the number of professional staff also increased four-fold, and he was the first to hire non-economic social scientists such as sociologists, anthropologists and political scientists.[2] In the years to come, however, what would become more important was his steady insistence that development should include a social and poverty orientation. He criticised the trickle-down concept, which assumes that

---

2   The numbers read as follows (Kapur et al 1997: 186): staff totalled 414 in 1948–49, rising to 1,859 in 1968–69. By the end of McNamara's time in office in 1980–81, 5,470 people were working for the Bank.

economic growth will automatically reduce poverty. He also argued against the main economic assumption at that time, which held that taking care of social policies goes hand in hand with negative trade-offs in growth. In other words, he was promoting both economic growth *and* poverty reduction. He turned the Bank from just a financial institution into a development agency.

However, he was struggling against the market-based background of the Bank and, at first, was met with little sympathy within the institution. He did have some outside support but, internally, his explicit and open focus on poverty alleviation was very much a personal vision and left him a rather solitary figure. In his first few years, he drew on his political experience and employed several strategies to upgrade poverty as an objective for the organisation. These included administrative changes such as organisational reforms and reorganisation, and the hiring of specific people for key positions. Most importantly, McNamara tried to exert influence through advocacy, using speeches, research publications, conversations with heads of governments and the like. Eventually, in the late 1970s, he pushed openly to integrate the basic needs approach (BNA) into the Bank's programmes, which would have highlighted social policies in bank lending to a much greater extent. However, many in the Bank were sceptical of this approach. McNamara required more research to support his position – but before anything could be done, this option was excluded by emerging economic instability and political conservatism, not to mention McNamara's retirement in 1981.

The late 1970s and early 1980s were a jarring time for the world economy, due to two major events in particular. On the one hand, the second oil crisis shook the world economy and led to extreme fiscal and monetary policies in the OECD countries, which in turn affected the international financial system. While the first oil crisis was addressed by increasing the resources available for non-oil-producing, developing countries, the second crisis provided leeway for adjustment policies. On the other hand, the Mexican debt crisis at the beginning of the 1980s, which was no mere domestic problem but had consequences that threatened the entire global financial system, provided grounds for a so-called "one-shot stabilisation adjustment" (Kapur et al 1997: 26). These events, combined with the success of conservative political parties in several key economies, led to a different type of economic policy and the resignation of Robert McNamara. He left the Bank in 1981, due to the fact that prevailing circumstances had all but eradicated the social perspective from the organisation's discourse. The notion that growth is the only engine for poverty reduction re-emerged, supported by the experiences of the Asian "Tiger" states (Hong Kong, Taiwan, Singapore and South Korea), as well as India, China, Pakistan and Indonesia, which recorded high rates of growth and poverty reduction.

This economic and political situation led to the introduction of structural adjustment lending (SAL), in 1979. The 1970s were economically unstable and, by the end of the decade, many developing countries were facing problems with debt and the balance of payments. For the Bank's lending programmes, this raised concerns regarding these countries' repayment prospects. Ernest Stern, by then (1977) McNamara's new operations chief, openly expressed his dissatisfaction with the Bank's project lending and its limited role in non-project lending, which he considered to be the reason for the failure to motivate developing countries to introduce macroeconomic policy reforms. McNamara endorsed Stern's proposal to exert greater influence on macroeconomic policies and expand the Bank's involvement in fast-disbursing lending not associated with poverty campaigns. At the Annual Meeting in Belgrade, in late 1979, McNamara announced the Bank's adoption of SALs.

The new SAL scheme was to fulfil two functions. On the one hand, it would employ lending in order to induce policy reform; on the other hand, it was a means by which to quickly deliver resources to developing countries. The latter function became highly important once the consequences of the second oil crisis took hold. However, it was not only demand from the developing countries that led the Bank to consider such a new lending scheme. In this regard, the organisation increasingly competed with other commercial banks. Commercial loans were extremely attractive to developing countries, given that they offered a low-cost alternative to Bank and IDA loans and imposed no extra policy-based terms and conditions. This happened at a time when the Bank was trying to expand its lending and influence, and it had a negative effect on the Bank's (political) leverage. The Bank was not necessarily in a position to play the envisioned role as a policy advisor – or, to put it another way, recipient countries were less receptive to its reform guidelines. This situation changed a couple of years later, with the onset of the Mexican debt crisis, after which the Bank gained influence and was able to exert it via conditions attached to SALs.

Once the concept of macroeconomic adjustment was incorporated into the debate, it increasingly displaced the focus on social or human development on the Bank's agenda. These developments coincided with the appointment of a new Bank president in 1981. Alden Clausen, an experienced, successful commercial banker, had little knowledge about either development or Washington. In addition, there were changes in other crucial positions – most significantly, the replacement of Hollis Chenery by Anne Krueger (vice president and head of the *Economics Research Staff*), a neoclassical trade economist who, in turn, initiated further replacements in the Bank. A new, neoliberal atmosphere became increasingly prevalent in the organisation and the formerly overt priority of addressing poverty, already in decline, completely disappeared.

External criticism of SAL continued to mount during the 1980s, mainly with regard to two different situations. Firstly, circumstances in Sub-Saharan Africa (SSA), in particular, deteriorated in the early 1980s, mainly due to drought and severe famine. The pictures of starving children in Africa that were shown around the world raised public awareness not only of the African crisis but also of the Bank's activities, both in this respect and in general. The Bank was criticised on the basis that, under such critical circumstances, SAL is not the most appropriate instrument with which to enhance economic and social development. The policy recommendations attached to such loans might even aggravate the situation of the most vulnerable. One of the most resounding voices behind this critique was UNICEF, which published a report called *Adjustment with a Human Face* (Cornia et al 1987).

The second source of external critique was NGOs. The early 1980s saw the Bank beginning to formally acknowledge NGOs. In 1982, the NGO-World Bank Committee was established – the first time that NGOs were institutionally integrated into the development process. In 1986, an NGO unit was created in order to monitor relations between the Bank and NGOs.[3] This was at a time when NGOs increased their campaigning against the Bank, in an attempt to generate public awareness about the environmental and social costs of Bank projects and to push the Bank towards greater openness and accountability. Three Bank projects, in particular, received a lot of publicity (Miller-Adams 1999: 7): the Northwest Regional Development Program in Brazil (1981–83); the Indonesia Transmigration (1976–86); and the Sardar Sarovar (Narmada) Dam in India (1985). The latter is particularly well known, as it led to the Bank conducting a comprehensive re-evaluation of its resettlement policies (see e.g. Fox 1998). After NGOs pushed the Bank to investigate the case in 1991, the Morse Commission revealed the Bank's non-compliance with its own policies on involuntary resettlement, which had been in place since the early 1980s (Operational Manual Statement (OMS) 2.30 and 2.33). This, in turn, caused a major internal review of the Bank's resettlement policies, in 1993–94, with the staggering result that most of the projects were found to have caused a huge displacement of people without any assistance. In the beginning, these NGO

---

3   It must always be kept in mind that the Bank does not negotiate projects or programmes with private stakeholders – or NGOs, for that matter – only with governments. Initiating more direct dialogue with NGOs, by creating special units inside the organisation and hiring staff who formerly worked for NGOs, was a big step. These developments continued in the 1990s with the formation of the Inspection Panel in 1993, to which private citizens who feel their lives have been or could be harmed by a Bank project can turn. For the first time, an administrative conduit was established for private individuals to make themselves heard by the Board of Executive Directors.

activities were only directed at specific projects, especially those involving issues such as the environment, indigenous peoples or women. However, their efforts had two main consequences. Firstly, they increased intrusive public scrutiny of Bank engagement in general, and, therefore, also in other areas of social affairs. Secondly, they eventually led to an increased focus on participation, and on people, in development.

As illustrated by the Sadar Sarovar projects, the internal advocates supported these developments. The Bank approved both the dam and the irrigation canals in 1985. While these projects were designed to bring drinking water to a population of 30 million, they also entailed the relocation of 140,000 people (Davis 2004).[4] The immediate response to the *Morse Commission Report* (Morse Commission 1992) was damage control regarding the specific project in India. However, an assessment was also required of all of the resettlement projects in the Bank's portfolio. Internal advocates initiated this second report, using the Narmada incident and NGO pressure to drive internal reforms. The *Resettlement and Development* report's findings were alarming: between 1986 and 1993, more than 2.5 million people were displaced by 192 projects.[5] An OED report, reviewing the lessons learned from the Narmada project, concluded that there was a need to ensure that projects were appraised by social scientists, sociologists and anthropologists, as well as economists and engineers (OED 1995). Two years later, Wolfensohn acknowledged this in his reorganisation of the Bank (the Strategic Compact, see below). The Narmada incident also led to the establishment of the Inspection Panel.

More generally, the engagement of several staff members in social issues predated the organisational activities of the late 1980s by many years. The tendency to be involved in social issues depended mainly on the profession of the individual in question. In 1973, under Warren Baum (then vice president of Bank Operations), a paper was circulated entitled *A Report With Recommenda-*

---

4   The number of farmers forced to resettle was 200,000 (Rich 2002: 27).
5   For more details, see Fox 1998: 313 or the report itself (World Bank 1994).

tions on the Use of Anthropology in Project Operations of the World Bank Group (Cochrane/Naronha 1973). Following the recommendations contained in this report, a small number of anthropologists and sociologists were hired in the mid-1970s. While these social experts were often employed for very practical purposes – such as translating or providing expertise on specific cultures/tribes – their influence on the overarching atmosphere should not be underestimated. At that time, the entire Bank was supposed to shift to a new development paradigm. McNamara created a special "pioneer division" headed by Leif Christofferson, who was given free rein to recruit a whole host of different people (interview with M. Cernea, 8 April 2004). Thus, the first cohort of a dozen or so NESSies *(Non-Economist Social Scientists)* entered the Bank and were scattered widely across the organisation. These developments had consequences.

Very soon afterwards, in 1977, a Sociology Group was initiated, led by Michael Cernea. This group did not have a formal status inside the organisation, and so had no direct influence on any decision-making process in the Bank. Rather, it was an informal community of like-minded people dispersed across the organisation. They met periodically to discuss specific social issues related to development, invited guest speakers from outside the Bank, etc. Its brown-bag lunch meetings and presentations were usually attended by 20 to 50 Bank staff (interviews on 7 and 8 April 2004; also Davis 2004; for more details on this group, see Kardam 1993).[6] Cernea reflects on this period: "Bringing social knowledge into the Bank was my challenge from the very first day I joined the World Bank, my 'ToR'[7] ..." (Cernea 2004: 3). The efforts of these "unorthodox staff members" were not merely restricted to attempts to discuss their perspective on development and convince others by writing papers and books. They also actively pursued policy changes within the organisation, mainly by

---

6   There were also other informal groups, such as the "Friday Morning Group" (Miller-Adams 1999).
7   "ToR" stands for "Terms of Reference", in other words, the job description set out when an employee first joins the Bank.

employing a strategy that they themselves called "getting from project to policy level" (two interviews on 8 April 2004). In other words, transforming specific practices deployed in a few projects, mostly regarding indigenous people, rural development and the like, into overall guidelines or directives that would then be applied to all Bank activities.

These internal advocates had even greater impact once James D. Wolfensohn came on board, as the World Bank's new president, in 1995. His time at the Bank is widely known as the period in which the organisation rediscovered poverty. On arrival, he found an organisation in crisis, with the Bank under pressure on several fronts. In addition to the increasing NGO critique, which became more and more public and culminated in the *50 Years is Enough* campaign in 1994 (see Danahar 1994), the organisation also needed to defend itself against another crucial criticism – of the "culture of loan approval". The 1992 *Wapenhans Report* revealed that a culture of loan approval in the Bank caused a decline in the performance and quality of its operations (Wapenhans et al 1992). The results were embarrassing: over one-third of completed Bank projects were failures; more than half of ongoing projects were assumed to be unsustainable; and less than one-quarter of projects in Sub-Saharan Africa were estimated to be sustainable, according to the organisation's own criteria. Increasing financial pressure on the Bank further complicated these deeply rooted organisational problems. The early 1990s saw an increase in the flow of private capital to developing countries,[8] which undermined the Bank's role and influence. This, combined with the growing perception of a gap between the Bank's rhetoric and its performance, caused a loss of faith in aid and development. Hence, Wolfensohn saw himself confronted with a number of challenges from a diverse range of angles. However, instead of choosing a few selected priorities, he wanted to serve all of these

8   Private capital increased from $40.9 billion in 1990 to $256 billion in 1997 (MIGA 1998: 5). In the same period, multi- and bilateral aid decreased from 57% to only 15% of total net development aid (World Bank 1998).

different interests. He was trying to be "all things to all people" (Rich 2002: 26) while also striving to ensure the Bank's continued relevance, influence and position of power. From his first day in office, Wolfensohn promised wide-ranging internal reforms in order to shift the Bank's culture of loan approval into a culture of "development effectiveness" and "accountability", with economic, social, and environmental policies as priority areas (World Bank 1995).

To that end, he proposed the Strategic Compact (SC), financed by a $250-million administrative budget increase over three years (from 1997 to 2000). This reform was supposed to renew the Bank's development paradigm and restore its external legitimacy and role as *the* leader in development. At the core of this reform were two items: firstly, a renewed understanding of development as a holistic process that needed to focus on social issues in parallel with economic growth. Secondly, the restructuring of the organisation, introducing a "matrix management system" geared towards reorganising the internal hierarchy, changing the incentive structure and therefore altering the Bank's culture.

The SC had a huge impact on the design of the social development approach. First of all, the social agenda was the core of this new Bank mission. In 1996, in conjunction with the Compact, a social development task force was set up.[9] Following the task force's recommendations, several institutional changes were implemented. In January 1997, the *Social Development Network* was established, situated within the vice presidency for *Environmentally and Socially Sustainable Development*. In March 1997, the SC was approved, and resources were allocated to the social development areas (e.g. for social analysis or to fund the implementation of regional action plans, as well as to hire more social scientists, in accordance with the task force's recommendation).[10] In October

---

9 It was at the urging of Michael Cernea that one of Wolfensohn's early acts was to request the formation of such a task force (interview with M. Cernea, 8 April 2004).

10 The SC reallocated US$10 million to the regions and US$2 million to ESSD for these issues.

1997, the Board approved procedures for adaptable lending, one of the task force's recommendations, in order to tailor operations more closely to each country's specific conditions. In December 1997, a research initiative was agreed upon to work on social issues together with the *Poverty Reduction and Economic Management Network* (PREM). By 1997, social development units had also been created in the regions, and a *Social Development Board*, including regional representatives, had been established. The board aimed to put in place an infrastructure that would make the network work, integrate it into Bank operations, make a number of topics mainstream (such as participation, social analysis and gender, identifying key social problems in each country and region,[11] incorporating social development into the Bank's business and working with NGOs) and focusing on new topics such as post-conflict reconstruction and cultural heritage.

The SC's organisational level reforms facilitated the Bank's ability to change. Combined with Wolfensohn's programmatic innovation – the *Comprehensive Development Framework* (CDF), in 1996, which reflected his vision on development, the organisation was already well prepared for a new development strategy focusing on social issues. Nevertheless, it was only after the East Asian financial crisis that the CDF became seriously operationalised, by means of the *Poverty Reduction Strategy Papers* (PRSP) initiative. The CDF basically reflected the president's vision of a new development strategy – yet it was not a great success, and never became policy itself. Instead, the PRSP initiative captured a few of the CDF's ideas, and served, so to speak, as the operational vehicle for the development vision. The East Asian crisis was, therefore, the catalyst for realising this vision. It provided a sound operational framework within which to implement a new concept of development, while at the same time drawing on the knowledge and foundations laid down after Wolfensohn's arrival, supported by a group of proponents of social development within the Bank

---

11 By 1997, for the first time in Bank history, each of the regions had a systematic action plan for social issues.

(interview on 8 April 2004). The PRSP initiative was launched in 1999 and was shortly thereafter adopted by both the Bank and the Fund.

CHAPTER 2

Paradigms and truth

The common perception of science is that it is cumulative, in other words that we are constantly becoming wiser and wiser as research breaks new ground. However, this view is not shared by everyone. The question of whether science is cumulative is of major significance for how statements are evaluated, and so criticisms of the cumulative concept have important implications for how we evaluate scientific knowledge and distinguish between knowledge *(logos)* and belief *(doxa)*.

The debate about whether science is cumulative has ebbed and flowed in both directions over the years. The Austrian philosopher and natural scientist Karl Popper (1920–1994) (see Chapter 3) posits that science is cumulative. He argues that evaluating the veracity of our hypotheses enables us to hone in on the truth about reality and to determine which statements are scientific and which are merely pseudo-scientific. In many ways, Popper represents mainstream understanding of the nature of scientific knowledge.

The cumulative concept is heavily criticised by the American historian of science, Thomas Kuhn (1922–1996). In his book *The Structure of Scientific Revolutions* (1962), Kuhn argues that science is not cumulative, but switches between different research paradigms – i.e. different sets of rules for what can and cannot be considered scientific knowledge. According to Kuhn, different paradigms have historically succeeded each other on a regular basis, in the sense that they cannot even communicate with each other or build on each other's results in order to generate more and better knowledge. They are what he labels *incommensurable*.

The English philosopher and argumentation theorist Stephen Toulmin (1922–2009) criticises Kuhn for his paradigm concept, and asserts that the history of science is evolutionary – in other words, the best and most durable arguments are capable of overcoming weaker ones. In an article called *History of Science and its Rational Reconstructions* (1970), the Hungarian-British philosopher of science Imre Lakatos (1922–1974) also challenges Kuhn's concept of incommensurability and argues that scientific progress is the result of constructive evolution, as part of which the best research programmes succeed each other.

The Austrian philosopher Paul Feyerabend's (1924–1994) book *Against Method: Outline of an Anarchistic Theory of Knowledge* (1975) takes the discussion of the cumulative understanding of science even further, with a continuation and radicalisation of Kuhn's view of science. He rejects the idea that it is possible to lay down credible and feasible rules to demarcate between science and pseudo-science, as illustrated by his famous quote: "Anything goes" (Feyerabend 1975: 28).

These debates underline that not all philosophers agree that knowledge is cumulative, and that we are continually building up an ever-greater understanding of the world and its phenomena. Views on cumulative science have huge implications for how we think that we can and should evaluate the results generated by our studies – not to mention how we actually do evaluate them in practice. This chapter deals with the challenge posed by the cumulative concept, later defences of a form of cumulative-science concept, and how we scientifically evaluate that something is true and something else is not.

Kuhn's history of science

Descriptions of perspectives in the philosophy of science often mention the concept of the paradigm. As a philosophical concept, the idea of the paradigm is introduced by Kuhn in *The Structure of Scientific Revolutions* (1962). The book discusses and challenges the generally accepted idea of science as a cumulative process that continually makes progress on the basis of methodological rules laid down by the science that is dominant at the time. Kuhn studied physics, and bases his description of science on the natural sciences. His intention was to describe what science has actually done in a historical perspective and what it is based on, rather than set standards for what it should be or do. His main argument is that a historical study of scientific change shows that science is not a cumulative process, but is based on abrupt breaks and crises. The development of science should be seen as continuous, almost cyclical. It moves from pre-science to "normal science" via a scientific crisis/revolution that leads to the (re-)establishment of a new "normal science", which in turn is replaced by a new scientific crisis/revolution and so forth.

Paradigms and normal science

According to Kuhn, a paradigm indicates both a worldview, i.e. an overarching shared idea of what the world is and can be, as well as a view of science (for a wider-ranging discussion of this, see Kuhn 1970: Foreword), which is based on universally acknowledged scientific positions that are exemplified by the problems and solutions presented by a research community over a certain period. This Kuhn describes as normal science. In his view, normal science follows a period of pre-science in which a large number of disparate theories and ideas about the world compete. In the pre-scientific period, none of the different approaches are raised to the canonical level. As a result, the nature of science and the question of what requirements to place on scientific work are constantly debated. This debate is held in what Kuhn calls normal science, which establishes an overarching conceptual framework and understanding in which some questions are plausible, while others are completely unthinkable. Normal science is thus a set of universally accepted values, rules and guidelines, all of which largely accept and set the frameworks for key questions, theories and methods. Another consequence is that debates and evaluations of scientific breakthroughs and discoveries are only held within this framework.

Kuhn lists various factors upon which normal science is based, and that help to ensure its continuation and the sense of community and affinity among its advocates (Kuhn 1970a: 181–191):

1. *Symbolic generalisations,* in which explicit knowledge is found and expressed. These can only be understood if we are familiar with the "cipher" used in the research – equations or curves plotted on graphs.
2. *Metaphysical ideas* about the nature of reality, which are not tested, but are taken as a general starting point for the research. In the philosophy of science, these ideas about the nature of the world are described as ontology (see Chapter 1). They are often expressed through models that determine what constitutes acceptable explanatory frameworks.
3. *Values and normative principles,* which form the background for deciding which criteria are applicable, for example, to a good theory, and can be used to make a choice between competing theories. They stretch beyond individual subjects and prescribe research values and norms that are valid for normal science.
4. *Exemplars:* Typically, a scientist, a textbook or a revolutionary article, which becomes an icon of good and proper research.

The role of normal science is not to challenge the paradigm but to maintain it, develop it and extend it to new and different areas. Kuhn calls research conducted within a normal-science paradigm *puzzle solving* (Kuhn 1970a: 38). He shows that science primarily develops and expands theories that are already widely accepted, and less often tests new theories that challenge the orthodoxy.

He points out that the role of the scientist is therefore to find and put into place all of the pieces that belong to the predefined area that makes up the puzzle. This means that both the problems formulated and their solutions belong within the paradigm and do not represent much of a challenge to it (Kuhn 1970a: 35–43).

Anomaly, crisis, revolution and incommensurability

According to Kuhn, even though it is difficult to challenge and go beyond the normal-science paradigm, paradigms are constantly exposed to *anomalies* (Kuhn 1970a: 52–66). These are phenomena that the paradigm and its theories are unable to explain or account for. Even if a series of anomalies occur, a paradigm will not immediately collapse. They are highly resilient and have their own internal explanatory logic. At first, it will typically be the incompetence of the individual researcher, rather than normal science, that will be criticised for the inability to answer or explain anomalous phenomena with the tools offered by the dominant paradigm.

However, the number of anomalies may be so great that it is impossible to accommodate them all within the current paradigm, which can ultimately lead to the paradigm breaking down (Kuhn 1970a: 97). Kuhn argues that when the normal-science paradigm collapses, a scientific *crisis* or *revolution* occurs. This period of crisis is replaced by a new normal science, which identifies new and different problems – among other things, it takes into account all of the unanswered and unexplained anomalies that have piled up and not been solved by the previous normal science.

Based on this review of history, Kuhn concludes that science is *not* cumulative and is not built on experience and knowledge of what is right and wrong accumulated under a previous paradigm. Rather, the new paradigm and the old paradigm are incommensurable (Kuhn 1970a: 122f).

Kuhn's thinking about paradigms and incommensurability represents a frontal assault on the cumulative concept. His thinking led to calls for a re-evaluation of how science is conducted and evaluated, and this gave rise to some of the ideas upon which constructivism (see Chapter 7) was based in the second half of the 20th century.

It is absolutely crucial that Kuhn was trained in physics and that the concept of the paradigm was conceived and developed in relation to the natural sciences. When we look at how Kuhn himself describes the paradigmatic developments, he identifies only a few fundamental scientific crises and describes the paradigms in very broad terms. As an example, Kuhn himself cites Nicolaus Copernicus (1473–1543), whose view of the world and of science – in which the Sun, not the Earth, is the centre of the solar system – eventually triumphed over the Ptolemaic worldview. Kuhn's idea that the different paradigms are incommensurable is an extension of this type of significant change of worldview in the natural sciences. To regard the Sun as the centre of the solar system was a radical change and could not be based on the normal science that preceded it, which presumed that the Earth was the centre of the universe. Using the concept of incommensurability, Kuhn is able to corroborate his rejection of the cumulative concept, and emphasises that it is radical changes – the results of crises and revolutions in views of science and of the world – that form the backdrop to paradigm shifts. For Kuhn, this is how science develops.

The paradigm concept in the humanities and social sciences

But can we apply the concepts of incommensurability and paradigm shifts to the social sciences and humanities? Kuhn's thinking has been criticised by various writers (Toulmin, Lakatos, Popper), and from the 1970s onwards debate raged about whether the same kind of paradigm shifts occur in the humanities and social sciences as Kuhn detected in the natural sciences. In the humanities and social sciences, it is possible to point out that ideas, symbolic generalisations, exemplars and values characterise that which at certain times might be classified as the main-stream. However, we cannot argue that the humanities and social sciences are under the sway of a single paradigm to the exclusion of all others. The humanities and social sciences bear greater resemblance to a pre-scientific stage, during which no paradigm is sufficiently strong that it eradicates the others completely. It also means that the different paradigms are not really incommensurable. The different theories and theoretical approaches are generally capable of understanding or speaking to each other if they so wish.

However, even if we are unable to identify pure paradigms or paradigm shifts in Kuhn's generalised and radical sense, it is still legitimate to speak of a kind of softer paradigm concept. Main-stream research and generally accepted theoretical perspectives exist in the various subjects and disciplines, as well as overarching mainstream research in, for example, the humanities and social sciences. To an extent, the mainstream is characterised by the elements that Kuhn uses to describe normal science.

How is the paradigm concept applied and what are its implications? Firstly, a "softer" version of the concept can be used, one that helps us to understand and describe how dominant theoretical approaches and perspectives in the philosophy of science attempt to block approaches with a different view of the world and of science.

Secondly, the concept can serve as the starting point for a discussion of the fact that science should not only be discussed *normatively* – that is, in an attempt to ascertain what science should be and what requirements you must live up to in order to be deemed scientific. Science can also be regarded as a historically changeable phenomenon, whose norms and values, in the same way as historical phenomena, are embedded in specific eras and contexts. This has given rise to a new discipline with a sociology-of-science purpose, i.e. to study how the creation of knowledge takes place. It is now a major and vibrant research area that incorporates both science and technology studies (STS) and actor-network theory (ANT).

Thirdly, the paradigm concept shows that the nature of science is not necessarily predicated on rigorous and universal rules. It is established via consensus among peers, which means that the same science concept and its evaluation criteria are upheld. This sense of community within a discipline does not just underpin and sustain eternal norms, it is also a social community that helps to set standards, maintain them for all newcomers and keep out those who do not adhere to the rules of the game/paradigm.

Fourthly, rejection of the cumulative and eternal evaluation of truth puts relativism on the agenda. If the nature of science is not based on universal rules and requirements that are applicable at all times and in all contexts, but depends instead on social communities, how do we determine that something is more true than something else? Indeed, is it feasible to talk about truth at all, or is it, as Feyerabend stated, a case of "Anything goes"?

Even though Kuhn's paradigm idea has spread relatively widely, and is used as an analytical concept with which to consider various elements of the philosophy of science, several other thinkers, e.g. Lakatos and Toulmin (see below), have countered Kuhn's idea of incommensurability and pointed out that knowledge can accumulate and that some arguments can be better then others, even if paradigms/research programmes are regularly replaced by new ones.

Lakatos and the understanding of science

The Hungarian-British philosopher Imre Lakatos weighed in with a heavy critique of Kuhn's concept of incommensurability without resorting to the naïve idea that social conditions have no impact on science. Like Kuhn, Lakatos came from a natural-science tradition. He paraphrases Kant in order to point out that it is impossible to separate accounts of the history of science from the philosophy of science: "Philosophy of science without history of science is empty; history of science without philosophy of science is blind" (Lakatos 1970: 91).

Lakatos criticises both the more naïve cumulative science concept and Kuhn's incommensurability. He emphasises that, in order to understand the history of research, you have to start with more general *research programmes* rather than singular hypotheses and theories. According to Lakatos, research programmes consist of a hard core of basic theories and assertions that form the basis for statements and theories within individual research projects, but that are rarely questioned by any of the scientists involved. If this inner core is challenged by either empirical or methodo-logical anomalies, a number of auxiliary hypotheses and theories are drawn up (negative heuristics) to protect the core or turn the anomalies into something positive that confirms the inner core (positive heuristics).

Thus, Lakatos believed that scientific development takes place through constructive evolution between competing research pro-grammes. In this scenario, the progressive and fastest-growing re-search programmes, i.e. those characterised by their ability to pre-dict and explain phenomena and events that have not emerged or happened yet, lead the way. Progressive research programmes are therefore at the forefront of empirical developments. By contrast, *degenerated research programmes* merely react to empirical evidence and attempt to explain away anomalies that challenge the pro-gramme. In time, most researchers will therefore be convinced by the strength of the progressive research programme.

The Toulmin model of argumentation

This evolutionary way of thinking is also evident in the English philosopher Stephen Toulmin's (1922–2009) understanding of science. Where Lakatos focuses on the more general research programmes, Toulmin thinks that the best and most coherent arguments "win out" over poorer ones, and that science makes progress when poorly underpinned theory is jettisoned in favour of theory with a solid foundation. Toulmin builds his discussion of paradigms on his work with truth, in the form of his theory of the valid argument. Where both Kuhn and Lakatos are interested in discussions of how scientific developments occur, Toulmin's argumentation model also serves as a practical model to either formulate valid arguments oneself or evaluate the validity of other people's arguments.

Toulmin's model of argumentation defines how arguments are and must be built up. According to this model, an argument consists of *grounds* that lead to a *claim*. He distinguishes between two levels of ground: direct ground, which directly indicates the evidence for a claim; and a more general ground, or the *warrant,* on the basis of which the direct grounds are chosen. *Backing* consists of justifications or empirical relationships that strengthen the claim. The more general justification, the warrant for the claim, is of a more general nature and serves as the link that ensures that the ground actually backs the claim. The warrant can be based on different types of generally accepted conditions, such as:

- generalisation (what is applicable to a small sample covers the greater whole)
- analogy (use of a case study to show that the same is true in another case)
- sign (where a sign is seen as an expression of a greater context)
- causality (that a given phenomenon is the result of or has arisen due to X)
- authority (reference to an authoritative source)
- principles (use of generally accepted principles).

It is not always easy to identify the warrant that is used to bind the claim and the ground together in an argument. It is rare for the warrant to be described in the text. It is mostly implicit. Therefore, in the light of the ground and the claim, we are able to work our way towards identifying the warrant used in a given argument. Ground and warrant work on many levels, and it is often the case that a warrant in one argument serves as a ground in another.

In this book, we are dealing with the warrant as something inherent in the characteristics of approaches to the philosophy of science, something that links the claim with a direct ground in the form of generally accepted philosophy of science approaches and methodology. Although it is rarely made explicit, we often use the implicit warrant that binds claim and ground together when we accept or reject scientific arguments. The identification of ground, claim and warrant is therefore a good tool for opening up the underlying philosophy of science assumptions and perspectives in our own and others' texts. Toulmin's basic model is shown in Figure 2.1.

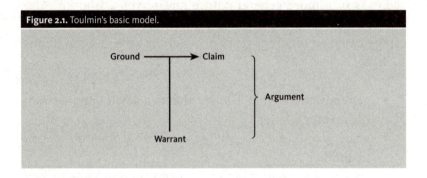

Figure 2.1. Toulmin's basic model.

Without prejudging the review of the philosophy of science positions in the chapters that follow, we can give a brief example. When a doctor identifies a link between smoking and cancer, the argument rests on the ground (direct ground) inherent in commonly accepted biochemical processes – if you do not accept these, the claim does not make much sense. However, at a deeper level the claim typically rests on the warrant (general ground) inherent in a *positivist* science perspective, according to which science consists of describing general laws on the basis of empirical observation.

Theories of truth

One of the basic problems raised by the paradigm idea is how we understand and evaluate the truth of scientific claims. The concept of truth has, of course, been discussed many times before.

The nature of truth and its consequences for knowledge and validity have been the subjects of lively debate all the way back to the Enlightenment. These discussions are precursors to the truth concepts that we today see reflected in the philosophy of science perspectives, all of which involve different ideas about what constitutes scientific truth. Like ontology and epistemology, truth concepts are a way of exploring and understanding the similarities and differences between different perspectives in the philosophy of science. In overarching terms, they are about the relationship between our scientific claims and that which we are seeking to describe. What criteria should we use to evaluate what is true and what is not? Are there degrees of truth? Can something be more true than something else? And if it can, how do we determine this? These issues are at the core of academic endeavours to evaluate both our own work and that of others. The truth concept is linked to epistemology – in other words, assumptions about how we understand the world and the material we deem useful for answering our questions. In order to evaluate what kind of knowledge will provide answers to our questions, we must start with what kind of relationship between the statement and the world must exist in order for the statement about the world to be regarded as true or credible. This book deals with the three most central of the many theories about the nature of truth: correspondence theory; coherence theory; and pragmatic truth theory. (For a more in-depth study, see Glanzberg 2009.)

Correspondence theory of truth

Correspondence theory stems from the tradition that further developed and reformulated what is known as *identity theory* (Russell 1910). Identity theory's dictum is that a true statement is identical to a fact. Critics of identity theory point out that the idea of identity is problematic, since statements and what they are about can never be said to be identical, because language and statements are not the same as the physical world. Advocates of the correspondence theory have slightly reformulated this idea by saying that a scientific statement is true if what is predicted *corresponds* with what the statement is about in empirical reality.

Specifically, correspondence theory means that we can and should test our statements' validity by examining whether they are consistent with the real world. Firstly, our statements about the world must be formulated so that it is possible to test whether they correspond with the world. Secondly, the reality in relation to which we are testing the statements must be accessible to us (for more on correspondence and representativeness, see Jackson 2006; Lynch 2009).

In research, the real world usually takes the form of numbers or percentages, which are considered to be condensates of the world. It is therefore also central for the validity of our arguments that we are able to show that these condensates, against which we are testing our propositions, can be said to represent the world. This also means that the choice of truth theory plays a significant role in study design, material collection and, not least, evaluating the validity of statements. Most scientific statements and theories contain claims about connections, relationships or causality. Empirically, these may be difficult to identify except via experiments. Unlike in the laboratory, it is rarely possible to conduct experiments in the social world, and therefore we have to demonstrate these links by proxy (indirectly), condensates or argumentation about probability. This often makes it difficult to adhere strictly to pure correspondence theory.

Criticism of correspondence theory

Pure correspondence theory assumes that it is possible to establish unhindered access to the empirical world, and that this will allow us to determine, once and for all, whether a statement is true. This assumption is found in the Polish mathematician and philosopher Alfred Tarski's (1902–1983) semantic theory of truth, in which he claims that the statement "the snow is white" is true if and only if the snow is white. This involves a 1:1 ratio between statement and world and only allows for the possibility that something is either true or untrue, which is obviously a problematic position to adopt. What if the snow is grey-white, muddy or brownish? Often, we have to simplify the description of the subtleties of the empirical world a little, which makes 1:1 correspondence between words and reality difficult.

Correspondence theory is also criticised on a more abstract level. Such criticisms stress that we do not have unhindered and direct access to the world, and therefore cannot test whether our statements are entirely consistent with it. We cannot empirically demonstrate everything in the world about which we would like to make statements. This has given rise to many philosophical discussions among both supporters and opponents of the theory. The result has been a softening of the correspondence theory, such that not all statements are said to correspond precisely 1:1 with that which they are saying something about.

Firstly, this has led to some writers seeing the correspondence theory as an ideal in which correspondence in practice can be established by proxy. This means that when a statement about, for example, a connection between two phenomena is difficult to prove empirically, it can instead be proved by the fact that the empirical statements do not contradict the hypothesis of this correlation – i.e. you do not prove the actual connection but what you can show is that the hypothesis does not conflict with the connection.

Secondly, some writers insist that our access to the world is never unaffected by our understanding of the world and our pre-conceived opinions. Kant's *das Ding für uns* emphasises exactly this. As a result, some writers do not see correspondence as an objective relationship between language and reality, but as an expression of how specific people see specific relationships in the world (see chapters 4 and 5).

Thirdly, some philosophy of science perspectives completely reject correspondence theory, arguing that language can in no way be said to correspond with what we are talking about, because the relationship between language and the world is essentially an expression of social conventions (more on this in chapters 7 and 8).

Fourthly, it has been pointed out that one problem with correspondence theory is that it is not clear what to correspond actually means. For example, can we say that the statement "the door is white" is true without explaining how we define "white" and "door"? Thus, rather than empirically ascertaining the validity of a proposition, correspondence theory opens up the prospect of infinite regression (for more on correspondence theory, see David 1994).

Coherence theory of truth

Another theory of truth is coherence theory. Coherence theory is all about *context* rather than agreement. A proposition is true if, in a non-contradictory manner, it is coherent with a set of propositions. What is central, therefore, is the proposition's relation to other propositions. The individual proposition is evaluated on the basis of its coherence with the whole set of propositions (the theory or empirical field) to which it relates. Therefore, we can also talk about degrees of truth in the coherence theory. A proposition may be more or less close to a set of true propositions, and we can talk about an overall understanding of the world. Propositions are tested not through their consistency with the empirical data, but by their coherence with other propositions about the empirical data and the world in general. Basically, this theory takes Kant's idea of *das Ding für uns* seriously – it stresses that we do not have unhindered access to the world, but that this should not prevent us from talking about it. We cannot say anything conclusively objective about the world, but we can evaluate whether our propositions are consistent with other propositions, and in this way we are able to approximate something resembling an overall understanding that does not contradict itself (see, for example, Walker 1989; Young 2001). The coherence theory of truth also opens up the possibility of commenting on phenomena that are not empirically observable, e.g. feelings, thoughts, ideas or contexts about which we cannot derive knowledge from direct empirical observation. Coherence theory can therefore lead to anti-realism, or at least a different way of thinking about reality other than the empirically based realism that forms the basis for correspondence theory.

Like correspondence theory, coherence theory has crucial importance for how we collect material, which material we use and how we evaluate its credibility and usefulness in our analyses. Unlike the empirical imperative in the correspondence theory, the coherence theory requires correspondence between the proposition and a set of propositions. Therefore, the coherence theory is typically used in philosophy of science approaches and theories that acknowledge that we do not have unhindered access to the world, and that reality is always mediated. Often, it is precisely these mediations of reality in which we are interested. Rather than identifying and studying objective conditions in the world, we are interested in individuals' subjective understandings of those conditions. Mediation, perceptions and understandings of the world cannot of course be tested objectively, but can only be tested in relation to the bigger system of mediations or perceptions of events in the society/group/person/era/texts/discourses with which we are presented. The validity and credibility of the individual proposition must be assessed in relation to whether it convincingly portrays the set of propositions to which it refers. Argumentation and probability are focal points for the coherence theory.

Criticism of coherence theory

The coherence theory has, of course, been criticised. Firstly, some applications of it have a tendency to use the same 1:1 ratio as the correspondence theory, only in this case in relation to the set of propositions (Glanzberg 2009). In these cases, the set of propositions is considered in the same way as reality is considered in the correspondence theory, and therefore the same problems occur.

A second, and weightier, criticism is of the anti-realism on which the coherence theory is based. If the propositions do not have to correspond with anything that can be identified as existing in the world, how can we champion one over the other(s)? How do we determine what is right and what is not? And how can something be partially right, or more right than something else? The theory allows for the coexistence of many truths, all of which are based on separate systems. As such, both the employees' perception of the organisation's oppressive ways of working *and* the management's understanding of itself as co-operative and accommodating can be true at the same time.

A third criticism is the difficulty of escaping what Kuhn called the normal science and creating new paradigms. As coherence theory is based on the idea that propositions must be coherent with a set of propositions, a scientific crisis and a paradigm shift are needed to replace the scientific system completely. We can therefore say that coherence theory is an extension of Kuhn's argument about non-cumulative science. Any replacement of a normal scientific truth system will not be built on top of the old system, but will rely on new contexts and sets of propositions.

Pragmatic theory of truth

According to the pragmatic theory of truth, something is true if it is useful or fruitful in practical or scientific life. Its focus is therefore on the *relation* between people, institutions and propositions. The pragmatic theory of truth was first defined by the American semiotician Charles Sanders Peirce (1839–1914), whose pragmatic maxims in the article *The Fixation of Belief* assert that "truth is the end of inquiry" (Peirce 1877). By this, he means that when something makes sense and helps to explain that which may at first glance appear inexplicable, it is true until you encounter something that makes you doubt what you previously believed was true. As a result, the truth process is not a closed unit, in which truth is processed once and for all. Rather, truth is continuously tested in relation to the specific context in which we find ourselves. According to Peirce's maxims, scientists must try to doubt their own assertions at all times, and they must then be verified again in "practical" use. The pragmatic theory does not deny that there is a truth, but asserts that the theoretical considerations must give way to what proves to be the case in practice. As the American philosopher and neo-pragmatist Richard Rorty (1931–2007) points out:

To say that we should drop the idea of truth as out there waiting to be discovered is not to say that we have discovered that, out there, there is no truth. It is to say that our purposes would be served best by ceasing to see truth as a deep matter, as a topic of philosophical interest, or 'true' as a term, which repays 'analysis'. (Rorty 2007: 17)

The consequence of the pragmatic theory of truth is that the results at which we arrive cannot be regarded as final and immutable. Our results must be the most useful and credible we can achieve based on the situation and the information we have at the given time and the context in which we formulate our results. Deciding how useful they are can, for example, mean that we deliberate whether our theories and conclusions describe and explain the phenomenon we are studying better than if we had not used them. To evaluate this, it is vital that we are prepared to continually expose our findings, theories and conclusions to doubt. We must try to find all the cases in which our theories do not make sense or do not contribute to a better and deeper understanding, and then review them and expose them to another, similar process. Credibility is established through the existing conventions for good science and the collation and processing of material. It is acknowledged that we cannot test the correspondence of our propositions with the phenomenon we are addressing, but this does not mean that there is no connection between them.

Criticism of pragmatic theory of truth

There are two main problems noted with the pragmatic theory of truth.

The first is that what is true can, in principle, change constantly. The truth does not have a fixed and unchangeable basis that can be secured and tested through agreement with the empirical evidence. Truth, or what is right, becomes a matter of judgement that places the individual researcher – rather than an objective scientific method – at the centre (as a form of compensation, the most convincing argument, the nature of which is of course objective, is central to the pragmatic approach).

Secondly, the pragmatic theory is closely related to the question of who or what is useful. This question relates to the ethics of research, which are challenged by the pragmatic approach, because no single type of utility and use is championed above all others. This means that the pragmatic theory could potentially lead to the strongest and most powerful voice deciding whether something is true or not.

This chapter serves as an introduction to the debates surrounding the issue of whether science is cumulative, constantly expanding on and qualifying our knowledge of the world. Kuhn's paradigm idea, and especially his concept of incommensurability, accuses this view of describing a desirable state that has no relation to real-life science in practice. The criticism raises the question of how we evaluate the validity of scientific results and confirm the truth of propositions. The chapter also presented three theories of truth: the correspondence theory, the coherence theory and the pragmatic theory. Which of these theories we use as the basis for scientific work is of critical significance for how we can argue scientifically in favour of the validity and credibility of propositions and analyses. As a result, it is also absolutely crucial when evaluating other people's results and putting the case for the credibility and validity of your own. Which theory of truth a particular philosophy of science perspective is based on will therefore be a recurring topic of discussion in the following chapters.

CHAPTER 3

Positivism and critical rationalism

Positivism is one of the oldest and most influential positions in the philosophy of science. Its close association with the professionalisation of most of the disciplines of today means that it plays a very central role in the philosophy of science. The term is often used indiscriminately, usually as a generic term for many different thinkers and philosophy of science perspectives from which the social and human sciences have attempted to distance themselves since the 1960s. In the Anglo-Saxon philosophy of science tradition, we often find a dualistic division in the social sciences between positivism and the interpretive sciences. In this context, positivism is characterised by its efforts to comply with the natural sciences' methodological requirements for objectivity and explanation. The problem with this division is, firstly, that it lumps the various aspects of positivism together. Secondly, a number of important ontological and epistemological differences between the various thinkers are confused and defined as positivist, so the distinctions become blurred. Thirdly, the similarities between what are classified as the positivist and interpretive sciences become unclear. Today, the term is usually applied to traditions that are based on logical positivism's empiricism, in which the credibility of hypotheses is evaluated through verification, and traditions that are based on critical rationalism, in which the credibility of hypotheses is evaluated with the help of falsification. The chapter introduces both traditions and emphasises the similarities and significant differences between verification and falsification.

Positivism and critical rationalism are similar in a number of key areas and are usually presented as a single unit. Consequently, we can make certain generalisations.

1. The view of science underpinning positivism and critical rationalism is *the unity of science*, which implies that science is seen as universal in nature. Accordingly, the same rigorous method can and must be followed to generate scientific knowledge, no matter which phenomena are being studied. This method is typically modelled on the quantitative-oriented natural science approach, which is proffered as an ideal.

2. According to Jürgen Habermas, the knowledge-constitutive interest is *technical*, since the purpose of the studies is *nomothetic*, i.e. to establish general laws or laws about *causal* connections through an *empirical-analytic* approach and thereby create the basis for predictions. In this way, the *explanation* of a phenomenon becomes the same as being able to describe the factors that led to the phenomenon occurring – or, as in functionalism, functioning as it does.

3. The ontological starting point is realistic. This means that positivism and critical rationalism are based on the assumption that the phenomena and causal connections in the study exist independently of the researcher. It is the researcher's task to identify them, by means of the rigorous scientific methods.

4. Empirical studies play a central role when considering epistemology. That which can be observed empirically exists in the world, while that which cannot be observed cannot be claimed with scientific credence to exist or be worth bothering about scientifically. The validity of scientific statements is established through observable empirical data. However, there is a difference between positivism and critical rationalism. Positivism takes an *inductive* starting point in the experience of sense-based *observations (a posteriori)*, which must subsequently be *verified*. Critical rationalism,

on the other hand, takes a *deductive* starting point based on logical hypotheses *(a priori)*, which must be *falsified* through *empirical observations*. The starting point for both perspectives is that they strive to achieve knowledge that is as *objective* as possible.

5. Positivism and critical rationalism are therefore based on the idea that science should be *value-free* and should not make ethical judgments or be used for political purposes. It is therefore crucial to be able to establish a demarcation between *logos*, that which is objective scientific knowledge, and *doxa*, which consists of political opinions.

6. In the positivist and critical rationalist tradition, the view of human nature is mainly characterised by *individual reduction*, which understands society as an aggregation of individuals. The starting point is the individual's psychology. The individual is understood to be rational and driven by what is rationally optimal for them (with the exception of the positivist tradition based on Durkheim's social facts).

7. The theory of truth employed in positivism and critical rationalism is the correspondence theory. Based on empirical observations as the starting point for either establishing or testing theories and hypotheses, empirical observations of the world are used to ensure the validity and credibility of our assertions.

The roots of positivism

Positivism is closely associated with the emergence and development of the scientific ideal. This dates all the way back to the Renaissance and its re-introduction of the Greek philosophers. The idea that knowledge is acquired via objective and empirical observation stems originally from Aristotle, who describes (e.g. in *Politica*) how knowledge is acquired through experience-based observations. Aristotle's observation-based epistemology stands in contrast to the written medieval tradition, in which the reliability and credibility of knowledge is assessed on the basis of the extent to which it is consistent with earlier writings. Interest in the empirically observable and natural world is seen in, for example, Leonardo da Vinci's works from the 1480s onwards, which seek to represent reality correctly, as it presents itself through sense observation.

The English philosopher Francis Bacon (1561–1626) introduced the experimental method, which involves the repetition of phenomena's responses to stimuli under the same conditions, thus enabling generalisations about the behaviour/emergence/reaction of phenomena. Generalisation would later become a hallmark of positivist research. These are all features that the humanist-inspired popes of the late Renaissance persisted with and set as requirements in order to legitimise the validity of the large amount of new knowledge that flowed into Europe as a result of the great explorers and the expansion of the world in all directions.

Although we are able to trace positivism's principles of truth and epistemology back to the ancient Greeks, it is only with Enlightenment thinkers such as Thomas Hobbes (1588–1679), John Locke (1632–1704) and David Hume (1711–1776) that the early Renaissance ideas are systematised and, together with developments in the natural sciences, start to define the idea of what science is and should be. One of the chief characteristics of the Enlightenment was the clash between religious dogma and the celebration of human reason. The Enlightenment philosophers were eager to systematise the world and spread knowledge based on reason, which they believed would help emancipate humankind from tradition and the yoke of habit. In his book *Leviathan* (1651), Hobbes explains how morality and ethics are conventions that have to be studied empirically in order to be described. Although Locke rejects Hobbes's view of conventional morality and ethics, they share an emphasis on empirical observation as the basis for science. In *An Essay Concerning Human Understanding* (1689), Locke describes sense experiences as the basis for knowledge, while Hume, in his three-volume *A Treatise of Human Nature* (1739–40), argues that social development must be described through empirical studies that would lead to the description of general causal laws. All three of these Enlightenment philosophers played an important role in the preliminary stages of the development of the concept of the nature of science, and how scientific knowledge is distinguishable from other forms of knowledge. Hume's effort to purge science of metaphysics led, for example, to the development of the verification principle. He states that a proposition about the world must be empirically verifiable via our sense experience in order for it to be valid. We cannot therefore say anything scientific about the world that cannot be verified empirically.

The introduction of positivism

Positivism was first introduced as a concept and a direction within the philosophy of science by the French philosopher Auguste Comte (1798–1857), who is known as the father of sociology due to his special interest in society as an object of study. Like the Enlightenment philosophers, Comte was keen to cleanse science of metaphysics. One of the points he makes in *Course on Positive Philosophy*, published in six volumes between 1830 and 1842, is that there is an evolutionary law governing how human knowledge must undergo a process of development from the theological understanding, via the metaphysical understanding, to the positive understanding. This law could be seen in the explanatory frameworks that humankind had historically used.

Comte outlines an evolutionary movement starting in ancient history, in which unexplained phenomena are considered an expression of supernatural intervention. From there, the process continues into the age of Christianity and the Church's dominance, where phenomena are explained in terms of religion. This stage differs from the earlier one in that the search for the ultimate cause is framed by religion and is considered an expression of a single abstract will. The positive stage, however, does away with the idea of being able to justify different phenomena on the basis of a single (the metaphysical stage) or multiple (the theological stage) underlying cause(s). The positive stage is the scientific era. In this era, according to Comte, the emphasis is on revealing the laws that govern phenomena. As such, the explanatory models change from justifying phenomena with reference to an underlying factor, to justifying the world via the laws that generate the existence of different phenomena and their behaviour. Comte believed that he lived in an era that was somewhere between the metaphysical and the positive stage, and that the purging of the metaphysical from science would accelerate the positive stage (Comte 1830).

Comte is often identified with this law, and therefore it seems easier to dismiss his ideas as an expression of the evolutionary period in which he lived. His achievements, however, are many and varied. He plays a major role not only in the development of sociology but also in social science research as a whole while the discipline was in its infancy. Comte's positivism establishes society as an object of study, with the natural sciences as the ideal. He regards all phenomena (natural, social and humanistic) to be ontologically identical, and in that respect continues, qua this ontological reduction, the unity-of-science principle. In his work to define how we can acquire reliable knowledge and explain society with the help of general laws, he links rationalism (reason) with empiricism (observation).

Comte's positivism is based on five key principles:

1. What really exists (non-speculative)
2. What is useful and beneficial
3. What is certain (not in any doubt)
4. What is accurate
5. What is constructive (not critical/destructive).

Even though several basic elements of Comte's positivism are reflected in modern post-positivism, differences also exist. In his principles – specifically, principles 1, 3 and 4 – observation based on sense experience is seen as the foundation of certain knowledge. They underline that knowledge cannot emerge speculatively and a priori, but must always be based on that of which we can be certain, and this surety is achieved through our senses. However, principles 2 and 5 reflect a normative position on what science can be used for. Science must not be for its own sake, but useful, constructive and edifying. According to Comte, then, there is a moral imperative to scientific work.

Science in someone's service was a common phenomenon in Comte's day, when the Western world was engaged in nation-building and many sciences emerged and/or were professionalised as national projects with the purpose of examining society, history and language at national level. As we will see, this separation still recurs in many subjects and disciplines. Whereas, pre-Comte, moral questions dealt, in particular, with what the individual ought to do, his moral imperative stems from societal thinking about how to foster socially moral individuals.

Experience via the senses and induction

The English philosopher, economist and political thinker John Stuart Mill (1806–1873) built on Comte's ideas and took positivism further. Like Comte, Mill was interested in how we acquire reliable knowledge about the world. He ascertains that the most reliable knowledge is derived from sense observation, which can be verified to ascertain indubitable facts. Thus, both Comte and Mill distance themselves from the German philosopher Georg Wilhelm Friedrich Hegel (1770–1831), one of the leading figures of German idealism. Hegel's starting point is that our knowledge is based on our ideas and thoughts about the world *a priori*, i.e. before our encounter with the world. Mill and Comte, on the other hand, claim that we only acquire reliable knowledge about the world *a posteriori*, in other words, based on our sense experience of the world. For them, science is based on sense experience, which can be systematically processed in order to formulate general laws and truths about the world.

In *A System of Logic* (1843), Mill identifies induction as *the* scientific method that should be used in all research in order to generalise from a limited number of sense experiences to general rules. The general truths and laws of scientific knowledge should be able to predict phenomena and subsequently control them. At the same time, these predictions can be tested through sense experience in order to verify whether the laws are right. This thinking is based on the principle of causality and is characteristic of positivism in all of its more recent and older forms. To predict phenomena on the basis of general laws or regularities, this approach takes as a starting point (expressed in a very simple equation) that, if a happens, then b will happen. This causality can obviously be multifaceted and gauged in degrees, but the basic principle is the guiding one for positivism.

Modern sociology and functionalism

Émile Durkheim (1858–1917) founded the first European department of sociology in 1895 and is known as one of the pioneers of the modern discipline. Durkheim's studies of society and social conditions were inspired by Comte, among others. His starting point is that even though an ontological difference exists between natural and social science, we have to use natural-science methodology if we wish to process ontologically different phenomena in a scientific manner. In other words, he went in for *methodological reduction*. In one of his main works, *The Rules of Sociological Methodology* (1895), Durkheim underlines his methodological starting point: "The first and fundamental rule [of sociology] is to consider social facts as things [...] a social fact is every way of acting which is capable of exercising an external constraint upon the individual" (Durkheim 1982: 13f).

This understanding of social facts made it possible to observe and describe social conditions as external phenomena that have an impact on the individual. Durkheim explains social phenomena by describing how they are one of many functions in a system.

As such, he is also one of the champions of functionalism, which was the predominant theoretical perspective from the 1940s to the 1960s.

Functionalism is usually explained on the basis of the organism perspective, in which each part can be understood in terms of the role it plays in enabling the organism to exist as it does. It is, therefore, not the existence of the actual social structures that is central to the functionalist, but the effect they have on the other parts of society and on human behaviour. By way of an example, Durkheim describes suicide rates in different societies, which can be observed and categorised on the basis of their relation to the individual society's institutions. In this way, suicide is explained not as the individual's own choice, but as an external social fact based on the way in which a society is organised, and its institutions, which exert influence on the individual human being's beliefs and understanding.

Logical positivism

The natural-science ideal is continued by the group known as the Vienna Circle. Its philosophy formulated many of the principles that characterise what we know today as logical positivism. Consisting of a group of Vienna-based scientists and philosophers, led by Moritz Schlick (1882–1936), the Vienna Circle discussed fundamental problems in logic, mathematics and physics. It took its starting point in the Austrian physicist and philosopher Ernst Mach's (1838–1916) intensive work to cleanse science of metaphysics. In 1929, Otto Neurath (1882–1936), Hans Hahn (1879–1934) and Rudolf Carnap (1891–1970) published *Wissenschaftliche Weltauffassung. Der Wiener Kreis (The Scientific Conception of the World: The Vienna Circle)*, which attempts to formulate a joint scientific programme.

Central to logical positivism is the discussion of how to maintain the requirement that all scientific propositions should stem from sense experience, while at the same time regarding the natural sciences (especially mathematics, with its theoretical formulations) as the scientific ideal. The problem is that the theoretical and logical language of mathematics (for example) does not meet the requirement that all knowledge must be based on empirical sensations. The logical positivists assert that for a proposition to be considered cognitively meaningful, i.e. able to express knowledge and be evaluated as true or false, it must be verifiable. Therefore, their work consists of conceiving of ways in which all kinds of sentences and statements can be verified either directly or indirectly by the senses.

By extension, the Vienna Circle develops the use of induction and verification. The inductive conclusion implies a shift from empirical cases to universal laws and rules. The requirement for verification is a continuation of the Enlightenment struggle against metaphysics in scientific discourse. In addition to induction and verification, the Vienna Circle and the logical positivists are known for taking issue with Comte's assertion that science should be useful and edifying. Rather, they feel that science should be distinguishable from normative statements and that it should be objective and value-neutral.

Positivism is generally based on empirical studies and makes conclusions inductively – in other words, it extrapolates from empirical cases to general rules or laws. In order to determine whether hypotheses, rules and laws are valid, it must be possible to verify all statements on the basis of studies of reality. This means that only those hypotheses that can be tested empirically, either directly or indirectly, may become the subject of scientific study. Karl Popper, in particular, studies the use of verification to test the validity of scientific statements.

Demarcation of scientific knowledge through falsification

The Austrian philosopher Karl Popper (1902–1994), who was close to the Vienna Circle and its thinking, was greatly interested in using certain demarcation criteria to solve the problem of distinguishing science from pseudo-science. This interest is illustrated by Popper's book *Logik der Forschung (The Logic of Scientific Discovery, 1935)*, which criticises the logical positivists' use of induction. In it, Popper asserts that it is impossible to make general statements based on conclusions drawn from empirical observations.

His primary argument against induction is that it is not logically possible to generalise universal statements from many singular experiences, even if they are repeated time after time. To illustrate this we can use the story of the inductive goose; every day from 25 December in one year to 23 December the next, a goose observes that a person approaches it at 10am every day for the purpose of providing food. However, the goose cannot expect this to continue as a general law for all eternity, because on 24 December the human will approach not to feed the goose, but to kill it and eat it for Christmas dinner. Therefore, we can never determine truth positively and definitively, as a situation or experience may occur that causes us to reject the current accepted truths. Secondly, Popper argues that all observations are theory-based. As such, it is not possible to arrive at scientific knowledge of the world on the basis of pure sense experience, bypassing any theoretical approach. The reason for this is that "Theories are nets cast to catch what we call 'the world': to rationalize, to explain, and to master it. We endeavour to make the mesh ever finer and finer" (Popper 1959: 59).

Instead, Popper advocates that science should be based on hypothetical-deductive inferences. A deductive inference differs from an inductive one by the fact that the formation of the systematic theoretical hypothesis takes place *before* we observe reality. As such, we move from theory to empirical data; from the general law to empirical cases. The validity of the theory/hypothesis is therefore tested on empirical conditions.

In contrast to the inductive method of reaching conclusions and verification as a validity criterion, Popper proposes falsification as the criterion for scientific demarcation. Falsification takes as its starting point the idea that it is impossible ever to determine truth and therefore only statements that can be falsified through empirical testing may be regarded as scientific. This means, first of all, that statements and hypotheses must be formulated in such a way that it is possible to reject them. A statement like "either it rains today or it does not rain today" must be regarded as pseudo-scientific because it cannot possibly be falsified. The same is true of statements that cannot be falsified empirically, such as the question of God's existence, the subconscious in psychoanalysis or Marx's laws about the economic basis of politics.

Secondly, statements can be divided into the weak and the strong. "Strong" refers to the statement's degree of generalisation and universalism. The more we can describe and explain the statement, the stronger it is. Strong statements therefore contain more information than weaker statements, even if they are later falsified. Typically, the more we know, the weaker and the more multi-faceted the statement with which we are able to work. This has consequences for the evaluation of knowledge. The best knowledge is that which is synthesised in the strongest possible statements that have *not yet* been falsified.

The induction problem

A classic example of induction is the problem of the white swans. On the basis of one hundred or one million observations of white swans, logical positivists would conclude that all swans are white. Critical rationalism, however, would point out that even if we look at one million white swans, it is impossible to conclude that all swans are white, only that many swans are white. In order to make sure that the statement about all swans being white is not false, we must always attempt to falsify it. This means that we must actively search for examples of a black swan. Each time that we try and fail to falsify a statement, it becomes stronger, because it becomes valid for a larger area. For example, perhaps we wish to test whether it is only in Denmark that all swans are white. If we study other European countries, our statement will not be falsified, and therefore it will appear stronger. If we then test our hypothesis in Australia and find one or more black swans, we have to reject the statement that all swans are white and instead say that, in Europe, all swans are white. The statement thus becomes more true but also less generally applicable.

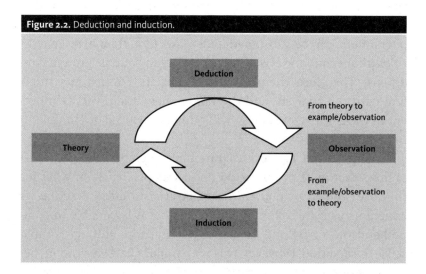

Figure 2.2. Deduction and induction.

Deduction

From theory to example/observation

Theory

Observation

From example/observation to theory

Induction

Establishment of the subject field in positivism and critical rationalism

What is it, specifically, that positivism and critical rationalism are particularly interested in studying and how is it connected to their ontology, anthropology and understanding of science?

Generally speaking, their knowledge-constitutive interest is in existing causal links. Causality involves a phenomenon being explained by a description of the factors that led to it occurring or, in a more functionalist approach (e.g. Durkheim), that the phenomenon works, emerges or is developed in a specific way. This can, for example, be the relationship between a new management strategy and an economic crisis, or it can be the different national cultures within an organisation and the quality of the work they do together, or the role of religion in medieval society.

The causal links are considered as general and universal, and

can be natural, social or specifically human. Positivism and critical rationalism adopt a unity-of-science view of science, which
means that all causal links can and should – both in principle
and on the abstract level – be studied in the same way, even if the
specific subjects are ontologically different. Explanations of individual phenomena will often consist of a description of how these
phenomena can be subsumed under laws (if a, then b) or general
rules (that economic crises impact on organisations' growth conditions). Therefore, the phenomena's specific context may inter-
fere with the analysis, and the randomness and diversity inherent in the context's meaning will often be reduced or eliminated
in the explanatory model.

In a positivist and critical-rationalist perspective, the causal links that exist in the world do so independently of whether
they are studied or not. Critics of positivism often describe its
adherents as *naïve realists*, who believe that it is possible to com-
prehend the world and its causal relationships through simple
sense observation. However, this is an unfair characterisation, because positivism and critical rationalism both use representative
realism, as existing phenomena are recognised through concepts
that serve to explain the general in events or to predict and control the future course of events.

In positivism and critical rationalism, the view of humanity
is characterised by a mechanical, atomic and typically behaviou-
ristic understanding of human behaviour. In the same way as the
causal process involves a cause and an effect, human beings are
seen as recipients of external stimuli, which almost automatically lead to a reaction. In this way, humankind is perceived of
as a mere object that is exposed to external or internal stimuli,
and whose actions are merely reactions to the different stimuli
to which they are exposed. This equation does not include the
complexity of social phenomena or changes that are not caused
by changes in stimuli.

The concept of the rational individual, which dominates
much economic thinking, especially in relation to constructing
economic models, belongs in this perspective. The rational indi-

vidual behaves as he or she does on the basis of what is best for the individual in a specific situation. This implies, firstly, that the individual is fully informed, i.e. is able to access all relevant information about the decision or action in question. Secondly, it requires that the individual actually behaves rationally and will ensure that all of his or her actions lead to the maximisation of utility.

With the exception of the Durkheim tradition, positivism and critical rationalism therefore take their starting point in *individual reduction*. This means that all collective phenomena and social conditions are studied and understood through individual actions. As Popper also emphasises: "[...] all social phenomena, especially the functioning of social institutions, should be understood as resulting from the decisions, etc., of human individuals, and [...] we should never be satisfied by explanations in terms of so-called 'collectives'" (Popper 1959: 98). We can therefore only understand collective and social conditions, such as an organisation's behaviour, by studying the aggregate behaviours of its members. We gain access to the whole (e.g. an organisation's culture) by putting all of the parts (the individual employees' cultures) together.

Examples of a research question

In relation to our case study of the World Bank's anti-poverty policy, this perspective would incorporate changes in external and internal influences into the formulation of particular questions about the causal relationships by studying how the World Bank's decisions on poverty reduction were affected by either internal or external political and economic conditions. The research questions could be, for example:

- What effect has the global economic crisis had on the WTO's approach to poverty reduction?
- What impact have different political institutions (e.g. the member states' governments, NGOs) had on the World Bank's anti-poverty work?

These questions would require insight into how the basis on which the World Bank makes decisions works in practice and how economic and/or political factors influence this. An obvious question would therefore be which general laws or rules about the relationship between economic/political conditions and the World Bank's anti-poverty work can be inferred on the basis of a study of these conditions over a prolonged period and in relation to different countries and priority areas. In a positivist-inspired approach, it is absolutely central to identify the relationship between cause and effect. Since it is possible to map the historical and current effects – in other words, to address and understand anti-poverty campaigns – the focus of the positivist study will be to identify the causes (i.e. the economic and political conditions) in relation to current and previous anti-poverty policies.

In this perspective, the relationship between cause and effect is considered to be causal and mechanical. The explanation of the relationship is based on more abstract, generally applicable and context-independent factors, e.g. economic fluctuations (including economic crises) that lead to changes such as the re-prioritisation of limited resources. This causal and general relationship takes place at all times, in all contexts, be it the ancient Greek city-states, China or Spain. In our example, the actual historical and cultural context does not influence the explanatory model.

Knowledge generation in a positivist and critical-rational study

A study of how causes and effects are interconnected must be conducted in a valid and reliable manner. In the positivist and critical-rationalist perspective, the ideal is natural-science methodology, in which there is distance between the researcher (the observer) and the research object that is being observed. The aim is to ensure the highest possible degree of value-freedom and objectivity. This is done by both the data acquisition and the analysis according with fixed methodological procedures that do not require continuous subjective interpretation, but which produce the same result when repeated.

Both positivism and critical rationalism incorporate observable empirical studies into the process of knowledge generation. However, they each have their own starting point. Where positivism bases its knowledge on an *inductive* use of empirical data, i.e. to say something true about the world a *posteriori*, critical rationalism, from a rational standpoint, takes as its starting point a rational theory that comes before the senses, i.e. *a priori*, and uses a *deductive* approach to determine whether the statement says something correct about the world. Positivism will therefore gather a certain amount of data, on the basis of which it will establish hypotheses about the relationships between phenomena. In a critical-rationalist approach, hypotheses are typically established on the basis of already formulated theories and existing literature, which gives rise to speculation that certain phenomena are causally related. This will subsequently be studied using carefully chosen material, so that the conclusion of the study can determine, in a valid and reliable manner, whether the hypotheses were correct or incorrect. As mentioned above, it is almost impossible to conduct pure inductive/deductive studies, but the different analyses' starting points are important in terms of the insight and knowledge they generate.

Both approaches require that the phenomena to be studied can be observed, and that causalities and functions can be identified and evaluated. In order to do this, qualitative elements are usually reduced to quantitative relationships, so that their mutual relationship can be measured and calculated. In order to ensure that the conclusions reached are reliable and valid, the empirical material is processed on the basis of a methodological unity-of-science understanding, with specific requirements for quantitative methodology. Therefore, it is essential that the research design for data acquisition ensures that the data is representative of the relationships we wish to study.

For example, it is very important to be able to say that the answers we may receive from questionnaire surveys can also be said to represent the group of individuals upon which we wish to comment. If not everybody can be asked – as is likely to be the case in studies involving very large populations – statistical methods are used to ensure that the results constitute representative, reliable statements about an entire group, even if we have only talked to a selection from the group. As individuals are thought to seek to maximise their own utility, and a collective relationship is always an aggregation of individual conditions, it is critical that the selection made is representative of the whole group. Depending on the nature of the study, the selection criteria can include, for example, seniority, gender, culture, age or education. The purpose is to ensure that we can evaluate whether the causal relationship at which we arrive through analysis is correct and consistent with the empirically observable reality. It is therefore about being able to argue for – and show that the conclusions correspond in a reliable and valid way to – the empirical observations.

Example of data acquisition and processing

What significance do the two questions outlined above have in relation to data acquisition? Consider the first research question: What effect has the global economic crisis had on the WTO's approach to poverty reduction? Here, we must be able to account for the causal connection. This means that we must be able to show/prove that the existence of, and the degree or number of, economic crises led to changes in the World Bank's poverty reduction. Material from 1970 until the present day forms the focus of our study. The development of the study design will require considerable knowledge of both empirical and theoretical factors. The economic factors we will look at are supposed to be quantifiable – or can at least be made so. We can therefore choose to study economic cycles by mapping economic conditions.

This can be done, for example, by looking at changes and fluctuations in total GDPs of the member states, international trade relations and global economic development. This data must then be compared with the development of the anti-poverty strategy. This can be described via the amount of funds to be distributed, the countries to which they are awarded or the projects deemed worthy of support. This allows the two elements of the relationship to be described. To answer this question, we also need to account for the effect of the initiatives. Here, we must consider carefully how any effect might be measured. In this context, we will perhaps opt for expert interviews with prominent people who have been central to the decision-making process and are familiar with the practical working procedures. They can explain how these decisions were made and give us an idea of the relationship between economic crises and the anti-poverty strategy. In addition to considerations of how we can measure the effect, we must also pay attention to factors that may act as distracting elements in the relationship. Expert interviews and existing literature can help us with this.

The relationships can be examined both inductively – i.e. through a study of all possible economic conditions and how they correspond with the change – or deductively, in which a clear hypothesis is established about the relationship and the hierarchy between the economic conditions, which is then tested against the relevant data. The design of the study and the data acquisition are important to whether the conclusions reached by a study of this type can be considered reliable and valid. This also includes the question of whether the conclusions and findings can be generalised and can confirm or refine existing general laws about how global and local economic factors influence international organisations. This evaluation is about whether the data acquisition, the data processing and the results can be said to correspond with reality to a sufficiently high degree.

CHAPTER 4

Hermeneutics

This chapter introduces one important reaction against positivism's definition of science, society and the individual – namely, hermeneutics. In hermeneutics, the starting point is that, rather than *explain*, as positivism does, we *interpret*. Phenomena are experienced by individuals who attribute meaning to them. This entails an ontological and epistemological concept of science that differentiates itself from the positivist ideal of the unity of science. While positivism wishes to explain the world once and for all, hermeneutics aims to account for how individuals perceive certain phenomena and events on the basis of a perspective rooted in a specific place and specific time. This introduces the contextual perspective, in which the focus is on the actors' understanding and intentions rather than specific events and phenomena.

Traditionally, hermeneutics has been closely related to the humanities. Its historical starting point should also be seen as a reaction to aspects of positivist dogma. We can make the following general statements about hermeneutics.

1. Hermeneutics represents a challenge to the *unity of science*. For the German philosopher Wilhelm Dilthey, the foundation of the hermeneutic perspective is that anything manmade is ontologically different from anything natural and therefore must be studied on the basis of quite different premises (see Chapter 1).

2. The subject field of hermeneutics is humankind, its culture and knowledge. Therefore, according to Jürgen Habermas (1968a), the knowledge-constitutive interest is practical, in the sense that science, through its focus on specifically human traits that must be understood and interpreted in their uniqueness *(ideographic)*, can provide a guide to human actions. The interpretation *(the hermeneutics)* of texts and phenomena are completely central to this form of science, which is oriented toward the individual's own understanding rather than objectively existing phenomena.
3. The ontological starting point for hermeneutics is realistic, i.e. it is assumed that the opinions and meanings studied exist in reality, and that it is the researcher's job to identify them. Opinions and meanings therefore exist independently of the researcher.
4. Hermeneutic epistemology is based on interpretation. Hermeneutics interprets text, speech or signs in order to analyse their hidden meaning and significance. The scientist's own prejudices and preconceptions form the starting point for the research, and this facilitates a better and deeper interpretation of the phenomena studied.
5. This starting point in prejudices and preconceptions means that hermeneutics does not consider science to be *value-free*.
6. Rather, the individual is considered to be intentional, i.e. he or she directs attention and actions towards something, and there is always an intention behind every action. Since interpretations always take place in time- and location-specific contexts, the meanings and opinions that the individual attributes to phenomena and events are considered to be contextual.
7. The theory of truth applied in hermeneutics is *the coherence theory*. In other words, a proposition is considered true if it is coherent with, and does not contradict, a set of interpreting propositions. It is therefore the evaluation of the relation to other interpretations and intentions that is central to evaluating the correctness and credibility of a herme-

neutic proposition (and not whether the proposition can be verified/resist falsification, as in positivism and critical rationalism).

The history of hermeneutics

The secret is in the text

Hermeneutics has deep historical roots in the Lutheran reading of the Bible, which, in contrast to Catholicism, tried to wrest the hidden meaning from the text itself (the Bible); this could only be done by interpretation. By making the Bible central to direct contact with God, the Lutheran clergy disregarded the Catholic priests' privileged status as special mediators of God's word. Hermeneutics therefore arose out of criticism of the Catholic Church's power in Europe. In this tradition, a text is seen as a sign of something else – initially, the word and meaning of God, which had to be wrested from the Bible's texts, descriptions and allegories.

In the 18th century, hermeneutics was expanded to include readings of all authoritative texts designed to provide a guide to human action. In extension of this, the 19th century German philosophers and philologists Friedrich August Wolf (1759–1824) and Friedrich Ast (1788–1841) developed a text-reading approach that they called the hermeneutic circle. They asserted that the basic principle of interpretation was that the spirit of the whole is found in its individual parts, and that all of the parts are found in the whole. We therefore achieve understanding of a whole through its parts – and knowing the whole allows us to understand the individual parts.

From text to author

The German Protestant theologian and philosopher Friedrich Schleiermacher (1768–1834) believed that hermeneutics, and therefore interpretation, is not only a way to understand and elicit the meaning from texts, but a universal human condition that is applied constantly, even when we read the newspaper, etc. In this way, Schleiermacher established hermeneutics as a universal theory of interpretation in general, and identified a distinction between its textual and psychological aspects. His theory therefore marked the change in the hermeneutic tradition, from revealing the meaning of the text on the textual level to the idea that the text's meaning represented the author's intention and should be understood in this light (Jensen 1986).

This way of thinking was further emphasised by Dilthey, who placed the text and the author's intention in a historical context. Dilthey considered human creative output to be an expression of significance and embedded meaning that could only be understood, but not explained, which was the objective of the natural-science approach. This understanding of the meaning and significance of human creative output could only be achieved through empathy and by recreating the context in which it emerged. Thus, Dilthey underlined that the particularly human aspect can only be understood by dint of the experience framework of which we as humans are a part – i.e. it is only because we know what it means to be hungry, in love, tired or other basic human emotions, that we are able to understand the meaning and significance of a human activity. According to Dilthey, we achieve understanding through emotional empathy and thereby recreate the experience of the author when he or she wrote the text or performed the action. As such, the human sciences deal with subjects that should be studied via methods that are fundamentally different to those employed in the natural sciences.

The difference between explanation and understanding forms the basis of Dilthey's emphasis on the ontological difference between the natural sciences and the human sciences, which makes it necessary to view them as distinct from each other. It is also for this reason that hermeneutics made the transition from a text-reading strategy to a scientific programme. Dilthey and Schleiermacher expanded the hermeneutic circle from being linked to the text, both as a whole and in terms of its constituent parts, to the text being understood via an understanding of the author, who in turn had to be understood via the time in which he or she lived. We can therefore talk of the dual character of interpretation: the text reflects not only the author's intention, but also the time and the circumstances by which the author is coloured. As a result, the era during which the author lives can be understood by reading individual works penned by contemporaries. Schleiermacher and Dilthey not only extended hermeneutics to take as its starting point the "author's" (performer or sender's) intention and the importance of the era for the human action; they also shifted from a narrow textual focus to describing all forms of human action.

From author to interpretation

Unlike traditional hermeneutics, the focus of which is how we interpret our world, modern existential hermeneutics – in particular, as described by the German philosophers Martin Heidegger (1889–1976) and Hans-Georg Gadamer (1900–2001) – is interested in how we interpret, and the role of interpretation in human action. This changes the emphasis from the relationship between the sender and the text, or between the sender and the era, to the relationship between what is to be understood and the individual who wishes to understand. It is often said that the difference between methodological and ontological hermeneutics is that methodological hermeneutics (Dilthey, Schleiermacher) looks at how people attribute meaning to specific conditions in the world in certain contexts, but insists that the reader is capable of viewing the world fairly objectively. Ontological hermeneutics (Heidegger, Gadamer), on the other hand, considers the contextual perspective to be a basic condition, which means that the reader's context also has to be incorporated. It is important to emphasise that this does not mean that the world does not exist independently of the interpreting consciousness. However, it does mean that we have to include the interpretive consciousness in our thinking.

Heidegger, who is also a central figure in phenomenology (see Chapter 5), identifies, via his concept of *Dasein* (being there) in the book *Sein und Zeit* (*Being and Time*, 1927), that humans are always thrown into situations and always base their approach to the world on personal preconceptions. Heidegger thus places the interpreter at the centre and asserts that the reading of a text is all about how we interpret it, and that the same conditions apply to all knowledge.

Heidegger's student, Gadamer, provides the basis for modern hermeneutics in *Warheit und Methode* (*Truth and Method*, 1986–1990). His starting point is that we only have access to the world through our interpretation of language and texts, which we can only understand through our preconceptions and prejudices. Taking Heidegger's thinking as his point of departure, Gadamer engages in a radical critique of positivism's attempt to purge scientific knowledge of preconceived opinions. He criticises the positivists' use of method and rules, which he regards as an expression of belief in the possibility of fixed routes to the objective truth about an object. He argues that our prejudices and preconceptions form the only basis for interpretation. This should not be seen as reluctance to countenance the new and different, but simply as a general condition concerning knowledge of the world around us. We cannot understand anything without a context in which to understand it. Therefore, when studying a phenomenon, we always start with our immediate understanding of it. By virtue of our tradition and experiences, we have an idea of what, for example, it means to be human or to participate in a hierarchical relationship or to be part of an organisation. These preconceptions are our only point of entry to the phenomena we seek to study and understand. However, we do not hold on to these preconceptions – rather, we let them be affected by the encounter with the phenomenon and, therefore, we reach understanding via a pendulum that constantly swings between our preconceptions and the phenomenon being studied. The latter influences our preconceptions until we achieve what Gadamer calls the *fusion of horizons*, where the interpreter's horizon merges with the horizon of the phenomenon being studied.

As an example, let us look at an interview with an employee

about his or her role in an organisational change. The phenomenon we want to understand is the employee's role. We have some preconceptions about this role, and it is these preconceptions that form the starting point for the questions we pose in the interview. The answers to our questions may cause us to change our preconceptions, as what the employee tells us may differ from what we expected or may add something we had not anticipated. This then becomes our new starting point (preconception) from which to pose the next question, the answer to which in turn means we have to change our preconceptions and pose a new question, etc. The fusion of horizons occurs when it is no longer the case that the answers we receive change our preconceptions. In other words, horizons have been fused and we have moved our preconceptions sufficiently to understand the phenomenon being studied.

Establishment of the subject field of hermeneutics

For hermeneutics, the subject field is humankind, its culture and knowledge. Humankind is a thinking, conscious and acting subject, and as a result human actions cannot be explained causally, but must be interpreted and understood in light of how those individuals experience and perceive themselves and their situation.

This does not mean that hermeneutics has no concept of explanation, just that it differs from a positivist understanding. In hermeneutics, explanations are a supplement to understanding and are closely associated with hermeneutics' view of humanity. Hermeneutics considers the individual to be an intentional being whose attention and actions are directed at something, and who acts with purpose. In this light, hermeneutical explanations can be divided up into genealogical or contextual.

- Genealogical explanations mean that human actions are presented by looking back at the historical context, and the previous experiences or intentions that serve as the catalyst for these actions. In other words, an action/event is explained with reference to its origins. For example, the development of the Scandinavian welfare states could be explained via path-dependency, i.e. the choices and actions we take are always limited and formed by previous ones, which in this case reflect the particular path of historical development that includes the Danish co-operative movement, Grundtvigianism and democracy since 1949. The concept of path-dependency is central here. It means that even though a development process has taken place in which past events set frameworks in which subsequent events are possible, their interrelationship is not causal and absolute, unlike in positivism.

- Contextual explanations explain human actions on the basis of the specific context in which they occur. This can include a description of the rules, institutions or cultural rationales that surround individuals and prompt them to act as they do. For example, certain institutional or organisational developments can be described with reference to their cultural roots.

Hermeneutics' starting point is realistic in the sense that the individuals' attribution of meaning and understanding of the world exists independently of the researcher's study. Therefore, hermeneutics is not interested in challenging and questioning the validity of the different meanings and opinions that people attach to events and actions. It is interested in understanding and describing them in detail and in depth, and perhaps also explaining them genealogically or contextually.

Where positivism focused in particular on things that could be seen and observed through sensory impressions, hermeneutics considers the intellectual and intentional to be the basis for everything. All actions and changes can be traced back to how people related to them in their thinking. Change is, therefore, not an external force, but is caused by an individual's understanding of external elements and the importance they are attributed in the specific context in which the individual finds him- or herself. Hermeneutics can adopt either a top-down or a bottom-up perspective. This means that the focus can be on the intentions and understandings of either high-ranking executives or individual employees. Hermeneutics accounts for situations and events in depth, and is therefore well suited to describing states, i.e. broader contexts that shape and influence events and actions. Developments and changes will typically be described on the basis of several consecutive states, as it is not enough to just describe how two distinct phenomena affect each other – rather, we account for how the phenomena and their effects are embedded in broader contexts or states that may change over time.

Example of a research question

In the case study of the World Bank's anti-poverty work, the knowledge-constitutive interest could be to understand how and why there were changes in the period 1970–2000. The shift could be perceived as an expression of the change in the understanding of social development and anti-poverty action over the same period. A potential research question could be: What changes took place in the World Bank's anti-poverty work during the period 1970–2000, and how did they relate to the understanding of social development in the same period?

Within the hermeneutic approach, the focus would be on identifying and examining the changes in order to understand where they come from. We would be interested in the actors who could exert influence on World Bank's decisions, i.e. politicians, executives and employees could be considered relevant actors – depending on how much weight we attribute to the individual executives' decisions, as well as the role we ascribe to employees or the national populations in decision-making processes. The change could, for example, be seen as an expression of the different motives and intentions for social development of successive executives in the World Bank. In this context, it could be relevant to study the difference between, for example, McNamara and

Clausen. This would reflect a genealogical approach, in which individual executives' motives, intentions and origins are central to the study.

Alternatively, we could look more specifically at the historical context. During this period, there was a financial crisis, the World Bank's poverty programmes did not work as intended and several of the member states had conservative governments. These factors affected how people thought about and made sense of anti-poverty initiatives. The focus therefore shifts from the individual executive and to a greater extent towards the context in which his or her decisions were taken. Here, it is not just the specific contemporary context that is central, but also the way in which it changes over time and the consequences of this for the individual's understanding of combating poverty.

Unlike positivism's interest in the general and universal, hermeneutics is interested in the *ideographic*, i.e. the particular and the unique. Each event and action is considered in its own specific context and has its own specific history and genealogy. Hermeneutics does not look for causal connections with universal application. However, this is not synonymous with the knowledge acquired via hermeneutics being anecdotal, and only interesting in relation to the individual case. In our example, an understanding of the change in the World Bank will not only make it possible to understand the specific change, but could also illuminate, on a more abstract level, the role played by individuals' opinions and intentions in how changes occur in big multinational organisations that have great significance both for nations and for individual people's living conditions.

Knowledge creation in hermeneutics

The text above describes the area of interest of the hermeneutic perspective and the assumptions implicit therein. Whichever phenomenon you choose to study has implications for the knowledge you think can provide answers to your questions. It also has significance for the material you work with, how it is collected and assessed.

Hermeneutics studies the individual's understanding and experience of actions and events. Therefore, problems raised in this perspective are illustrated through material that enables access to the individual's thoughts, understandings and attributions of meaning, both in the present and in the past. It follows that material collected through the researcher's sensory observation is not directly usable. Often, it is said that strictly quantitative material, such as that used in statistical studies, is unsuitable. However, in this context, it is important to stress that hermeneutic approaches can also have a quantitative dimension, in the sense that it is important to have enough material to be able to achieve a full and complete understanding of the phenomena. The difference is more to do with the nature of the material in which you are interested. Knowledge that is relevant in a hermeneutic perspective is difficult to acquire by external observation of individuals, i.e. by separating the material from the actors about whom it says something. Rather, material is required that provides access to the individuals' understanding and intentions. Typically, this knowledge will be garnered from reading texts that, directly or indirectly, describe perceptions of incidents that have occurred, or by talking to the people who were involved in the incident and asking them to reflect on it in an interview.

A number of different methodological approaches have been developed that help to extract information from material that does not directly represent the individuals' experiences and understandings. Overall, we are able to posit that it is language, both written and spoken, that conveys the individual's inner attribution of meaning to others. That is why texts, in the broad sense, are absolutely central in the hermeneutic perspective. Hermeneutics interprets text, speech or signs in order to analyse the meaning and significance concealed in them. Its starting point is that the researcher's preconceptions about the phenomena provide access to a better and deeper interpretation. The interpretation can be characterised as subjective, since the knowledge about the phenomena being studied cannot be obtained by separating the interpreter from the interpreted. The interpretation is successful only when the researcher's own assumptions and prejudices are challenged. As such, knowledge in this perspective is seen not to be objective and value-free.

Hermeneutics works both deductively – individual phenomena and events are described by their function in the context (i.e. the whole explains the parts) – and inductively – the context is established on the basis of small traces and various propositions that are pieced together (the sum of the parts equals the whole). The careful and detailed description of a phenomenon and its context is the hallmark of these perspectives. A proposition is true if, in a non-contradictory manner, it is coherent with a set of propositions. This means that the conclusions you reach must not conflict with the detailed description on which the proposition is based. The quality and quantity of the material you collect in order to say anything credible about a phenomenon must therefore be sufficient to ensure that the collection of additional material will not significantly conflict with the proposition.

Example of collecting and processing material

In our case study (the World Bank's anti-poverty measures), the inclusion of both texts and interviews could lead to a broader and deeper understanding of the contexts, individual intentions and meanings that help to bring about the change. First, we must select documents that shed light on your problem and select individuals with whom it would be appropriate to talk. When making this selection, it is extremely important to know the decision-making process in the World Bank with regard to anti-poverty measures. It is clear that it is a prolonged process that, at any given time, involves external knowledge, discussion between individual members, the employees' execution of decisions and the President of the World Bank. Relevant documents from the 1970s to 2000 could be used to illustrate how decisions were made, who was involved and what arguments were used (and therefore considered to be legitimate). Interviews would have to be carried out with selected employees who either worked for the World Bank during the period in question or held other key positions, and are therefore able to talk about or represent changing viewpoints regarding anti-poverty action.

The scientist having a preconceived idea of what the change entailed and meant would reflect the hermeneutical spiral as a methodical tool. This preconception would be qualified through a number of interviews, which would lead to greater understanding of the various nuances in the change, as well as how the change was seen by employees of the organisation, and the consequences of these perceptions for the fight against poverty. In this way, the researcher's horizon would, by virtue of their changing preconceptions, constantly be moving until they achieved a fusion of horizons with the knowledge contributed by the interviewees.

Depending on whether the study had a top-down or bottom-up perspective, the focus would be on the intentions and actions of either the World Bank's presidents or the individual employees. Access to these could be gained through interviews with the parties involved, their own descriptions of their lives or others' credible evaluations of the intentions that lay behind the main actors' tangible actions. For example, a top-down perspective could focus on the role played by the varying backgrounds of successive presidents and their understanding of, and intentions regarding, social development. A hermeneutically inspired problem could focus, in particular, on describing in detail the knowledge and conceptual context in which the World Bank's anti-poverty work was carried out, the intentions of the people involved in these actions and their understanding of combating poverty. This could involve both a deductive and an inductive work process:

- Using a more deductive approach (in which we move from the whole to the part), the change in anti-poverty work during the three decades 1970–2000 can be understood through changes in the ideological context. We would therefore focus on a detailed description not only of the context but also of how it affected the central actors' actions.

- A more inductive approach (in which we move from the part to the whole) could be based on how each of the key actors' intentions and actions could be seen as a reflection of the development of the conceptual context.

CHAPTER 5

Phenomenology

This chapter introduces phenomenology. Phenomenology is one of the dominant directions of 20th-century philosophy and has assumed increasing importance in the humanities and social sciences in recent decades (Zahavi 2003). In particular, phenomenological methodology has been debated and posited as a response and alternative to the more structured and controlled approaches of qualitative and quantitative analyses.

However, phenomenology is not only a methodology; it is also based on a philosophical approach that is both similar to, and different from, hermeneutics, with which it has been interwoven since the beginning of the 20th century. The purpose of both is to understand human experience and action, but where phenomenology is interested in describing and understanding phenomena and practices, hermeneutics concentrates on describing and understanding interpretations. Hermeneutics seeks to identify the meaning in texts, the individuals' perceptions of events and the intentions behind the actions taken by individuals. This is done by taking the researcher's preconceptions as the starting point. The phenomenological perspective focuses on how the world (events and phenomena) manifests itself to the individual, without the researcher's preconceptions regarding the phenomenon interfering with this understanding.

Overall, we can summarise phenomenology as follows:

1. Phenomenology, like hermeneutics, challenges the concept of *the unity of science* – and especially its requirement for methodological reduction, which is based on an assumption of a distance between the observer and the observed. A natural-science approach separates the subject (the researcher) who does the observing, from the object (the topic) being studied. Phenomenology challenges this dualism and multiple other basic dualisms that pervade Western philosophy.

2. Phenomenology's subject field is human consciousness and knowledge. The phenomenological perspective assumes that phenomena are always phenomena for someone, and can therefore never be studied independently of how they appear to a particular consciousness. As such, the subject of phenomenology is the way in which phenomena manifest themselves to human beings. Since phenomena manifest themselves differently to different people in different contexts, the phenomena must be understood and interpreted *ideographically*, in relation to how different phenomena manifest themselves to different individuals.

3. Overall, phenomenology is also an attempt to abolish the dualism between ontology and epistemology. Objects do not exist in a vacuum, but in the different ways in which they appear in the world (phenomenology stems from the Greek verb "fainesthai", which means to show or present yourself). The objects only become something (ontology), when there is someone who knows them (epistemology). The world does not exist, it manifests itself, and cannot therefore be separated from the context in which the subject experiences it. The ontological starting point can be formulated as "the phenomenon's essence is its existence". This means that phenomenology is neither materialistic nor idealistic – but both at once.

4. Phenomenology is about epistemology, i.e. about how we know the world. In phenomenology, the interpretation starts with the phenomenon and cannot be separated from

how phenomena are and can be experienced. Therefore, from a phenomenological perspective, it is not possible to separate the subject from the world. The world cannot be considered objectively because there is no special, privileged position from which to observe an object. Any observation is always historically, culturally and experientially embedded.

5. Phenomenology is therefore based on the idea that science can never be *value-free*.
6. The phenomenological perspective perceives the individual as intentional – someone who acts with intention, who directs his or her attention and actions towards something. In doing so, it takes as its starting point the individual's intentions and intentionality, which also form the basis for understanding the nature of phenomena.
7. Phenomenology's underlying truth theory is – like hermeneutics – the coherence theory, i.e. the truth is arrived at by ascertaining that it does not contradict the overall understanding of what is being studied. It is important that phenomenology's descriptions of the world correspond to the people who live in it, even if the academic descriptions formulated are more abstract than everyday language. Credibility is evaluated in relation to the area described.

Early phenomenology

Phenomenology is far from being a homogeneous movement. It counts among its advocates great philosophers such as Edmund Husserl (1859–1938), Martin Heidegger (1889–1979), Jean-Paul Sartre (1905–1980), Maurice Merleau-Ponty (1908–1961) and Emmanuel Lévinas (1906–1995), to name but a few. The subjects addressed range widely, and phenomenology has had a major impact on the humanities, social sciences and natural sciences.

The German philosopher Edmund Husserl (1859–1938) is considered to be the founder of modern phenomenology. It is in his writings that many of the concepts that are still considered central were first formulated. Husserl studied different types of conscious experiences and their relation to intentionality. In this, he derived inspiration from the American pragmatist William James and the German psychologist Franz Brentano. Husserl's basic premise is that human experience is always directed towards something, and that this something can only be experienced by a subject – Husserl says that consciousness always has an *intention*. A phenomenon is therefore always something for someone, it does not exist in and of itself. Consequently, Husserl rejects Immanuel Kant's division of phenomena in the world into *das Ding an sich* and *das Ding für uns* that characterises many other philosophy of science perspectives. He takes as his starting point the approach that a phenomenon do not have an inherent essence.

In his book *Die Krisis der Europäischen Wissenschaften und die Transzendentale Phänomenologien (Crisis of European Sciences and Transcendental Phenomenology,* 1936), Husserl describes how phenomenology studies different structures of experience – thoughts, memories, perceptions, ideas, feelings, desires, bodily sensations, etc. – all of which involve intentionality. His intentionality concept is described in, among other works, *Logische Untersuchungen (Logical Investigations,* 1900–1901). It is based on the view that meaning emerges from human experience (i.e. that something makes sense for us) and is always based on consciousness being directed towards what it experiences (Zahavi 2003). This means that the discussion about the outer world's existence is no longer relevant, as phenomena have to be understood in terms of human experience *(lifeworld)* and therefore must necessarily be based on a *first-person perspective*. It is this subjective perspective that facilitates and conditions how reality manifests itself.

Husserl's phenomenology totally rejects the traditional scientific objectification of phenomena, in which the sentient subject's perception and experience of a phenomenon is considered biased. Consequently, we must try to purge the phenomenon of subjective bias in order for the clear and correct object to emerge. With his maxim "Zurück zu den Sachen selbst" (back to things in themselves), Husserl stresses the importance of understanding phenomena as they manifest themselves. Instead of getting lost in speculation and theories, we must take as our starting point the view that the world is given before it is theorised and conceptualised. Experiences must define the theories – not the opposite. This means that Husserl's work and phenomenology in general is markedly inductive (i.e. theory is formulated on the basis of empirical data, rather than theories being tested empirically). The inductive means of arriving at conclusions is characteristic of phenomenology in general.

Husserl insisted that it was the positivist perspective that did not give a clear and true picture of reality and added the methodological requirement that, unlike positivism's third-person perspective, facilitated a first-person perspective and an understanding of how people produce knowledge about their social reality. In a first-person perspective, the researcher must put aside his or her theoretical knowledge and question any preconceived ideas he or she may have in order to observe how phenomena manifest themselves. This, he called phenomenological reduction, bracketing or *epoché*.

The role of interpretation: Essence is existence

The German philosopher Martin Heidegger (1889–1979) plays a major role both in hermeneutics and in the further development of Husserl's phenomenology. His book *Sein und Zeit (Being and Time*, 1927) aims to describe the conditions for human knowledge and understanding. Rather than consider knowledge as something that arises only within the individual, Heidegger perceives the human being as a "being-in-the-world" *(Dasein)*, a world in which all meaning that arises from human experience must necessarily be interpreted through language, because it is the medium that people use to endow experiences with meaning. Human culture and lifeworlds are therefore inevitably expressed via language, which, consequently, can be used to access and understand human reality and experiences. Since all understanding is mediated and requires a specific world horizon (Heidegger 1927: 148) – i.e. a certain subjective standpoint from which to understand the world – Heidegger states that we cannot speak of essence, but only of existence. Things are the subjective knowledge of them. Heidegger also emphasises the object's own way of manifesting itself. He asserts that there is something special about the object's way of being in the world that is visible in its various manifestations. Therefore, all of the ways in which a phenomenon manifests itself are relevant to understanding the phenomenon in depth. The point here is that the phenomenon we endeavour to understand consists of the interaction between its specific manifestations *and* the structures of understanding that allow this phenomenon to manifest itself in the way it does to the interpreting consciousness (Heidegger 1927: § 7).

The presence of the body

In *Phénoménologie de la Perception (Phenomenology of Perception,* 1945), the French philosopher Maurice Merleau-Ponty (1908–1961) further develops Husserl's concept of intentionality – the idea that consciousness is always directed towards something. Merleau-Ponty's interpretation expands the concept of intentionality by making the body the centre for experiences. Thus, he expands Husserl and Heidegger's more language-based understanding of intentions and points out that intentions are about how we physically move around in the world. Like the earlier American pragmatists, he considers bodily sensory experience to be the basis for knowledge and understanding.

According to Merleau-Ponty, we cannot understand or describe the human being as separated into body and soul. On the contrary, the individual is a corporeal being-in-the-world. He therefore further underlines phenomenology's dictum of throwing off science's distinction between body and soul, and between object and subject. Unlike positivism, which distinguishes between the internal (subjective) and the external (objective), Merleau-Ponty and phenomenology insist that the true and proper understanding of phenomena is found in the form in which they manifest themselves in the situations in which we describe them. Merleau-Ponty also deviates from the more existential phenomenalism formulated by Heidegger (and later Sartre) by focusing on the interpretation of forms of manifestation as context-dependent and on the idea that words and social conditions derive meaning from the specific living conditions under which they occur. This underlines the importance of understanding a phenomenon's existence in a specific time and place in order to be able to understand its essence.

Establishment of the subject field in phenomenology

Like hermeneutics, phenomenology challenges the concept of *the unity of science*, and especially its requirement for methodological reduction, which is based on distance between the knower and the known, because the phenomenological perspective is based on phenomena always being phenomena *for* someone, they can never be studied independently of their manifestations. Phenomenology's subject field is human consciousness and knowledge – more precisely, the ways in which phenomena manifest themselves to humans and in human consciousness. Since phenomena can manifest themselves differently to different people in different contexts, the phenomena must be understood and interpreted *ideographically*, in relation to how different phenomena manifest themselves to different individuals.

Other perspectives, such as positivism, make a virtue of separating the object from the way in which it manifests itself. Phenomenology challenges precisely this, pointing out that the phenomenon can only be known when it manifests itself.

As such, we cannot understand an event objectively by disregarding our experiences of it. For phenomenology, to see the event as part of a causal chain that describes various successive situations, or to understand phenomena by reducing them to elements of an abstract theory (as is often the case in, e.g. psychology, sociology or biology), does not pave the way for understanding the event – on the contrary, approaches of this kind draw attention away from the actual event itself. The "objective" approach aims at removing our presence, doings, experiences and feelings that we associate with the event, to present it without distortion. However, all of these things are part of the events. We can never describe something properly if we do not start with how it is perceived and experienced by the people involved, from a first-person perspective.

The ontological assumption is that the essence of a phenomenon is its existence. This means that phenomenology is neither materialistic nor idealistic, but both, inseparably – the phenomenon's being (idealism) is simply its manifestation (materialism). The manifestation of the phenomenon is not indicative of a hidden core that controls events. This means that phenomena studied in a phenomenological perspective must always be understood as they appear via human experience.

In this context, the individual's lifeworld plays a major role. Lifeworld is an umbrella term for the subjects' horizons – their plans, purpose, intentions and dreams in the world – and constitutes the starting point for knowledge and understanding. The lifeworld is people's everyday life and practices, their mutual interrelationships and experiences, which create the background for their intentions (directedness) and the conditions for the way in which phenomena manifest. The phenomenological view of humanity sees the individual as intentional – someone who acts with intention, who directs their attention and actions towards something. For a research process, this means that we take as our starting point the individual's intentions and directedness, and use them as a basis on which to understand the nature of phenomena. Phenomenology's contribution is to point out that the subject cannot be separated from the world, because the subject's physical, social and cultural way of being in the world determines what the world is capable of being.

Examples of a research question

In adopting a phenomenological perspective towards our case study of the change in anti-poverty work at the World Bank, we must choose between different focal points. Typically, the object of our attention is defined in a micro-perspective based on the individual. It uses his or her perspective to examine the case without reducing it to an expression of a larger entity, such as society, the economy or similar categories. The research question is therefore directed at what the change in anti-poverty strategy actually was/is, from the perspective of the individuals who were affected by it or involved in the change. How do the changes manifest themselves to different individuals involved, e.g. the recipients, the administrators and those who formulated the different strategies?

A description of all aspects of the change from the 1970s to the 2000s would be overwhelming in this perspective, so we typically concentrate on an illustrative case that shows how practice, experiences and intentions interact in the lifeworld concerned. Here, you could look at everyday practices, routines, habits and patterns associated with the change before, during and after the famine in the South Sahara, focusing on the people involved, such as those who formulated or managed an economic strategy or those who received the aid. The lifeworld that these people inhabited would be central in revealing their experiences of what happened, and their intentions.

Knowledge creation in phenomenology

Phenomenology is about epistemology, i.e. about how we understand the world. The interpretation is embedded at phenomenon level – it cannot be separated from how phenomena are experienced or what they actually are. From a phenomenological perspective, it is not possible to separate the subject from the world, because the world can only be understood through the subject's interpretation. Phenomenology is, therefore, based on the idea that science can never be *value-free*. Further, it is precisely the attribution of value that constitutes the manifestation of the phenomena. Phenomenology does not seek to explain why phenomena become visible. The focus is on what phenomena are. Natural science is based on the relationship between a subject (the researcher) who considers an object (the object of the research). Phenomenology's aim is to do away with this dualism. It believes that research takes place between a subject (the researcher) who observes another subject's understanding (the object of the research) (Christensen 2002: 141). Therefore, it is absolutely central that the phenomena are understood through the subjects' experiences and actions in a specific context. The maxim "back to things in themselves" implies that we must always understand actions and phenomena in relation to their lifeworld. Phenomenological analyses are therefore always highly specific, descriptive and context-based.

The purpose of a phenomenological analysis is not to understand something new in the world, but to understand basic conditions in the world that are not always noticed and that are considered self-evident. This has methodological consequences. Husserl prescribes that we adopt a reflexive bracketing, the so-called *epoché*. This is a key analytical aspect of phenomenology, which requires that a researcher, from the outset, relates to the subjects' description of all phenomena as if they were completely unfamiliar. All prejudices must be put aside in order to really understand phenomena from a first-person perspective.

It also requires a different relationship to the collation of material. Initially, we merely collect in-depth and exhaustive amounts of data without reducing it to something else or eliminating the material that is not immediately comprehensible. The material is only analysed and interpreted after this open collection process. The reason for this dual approach is that if you start to acquire data on the basis of a very specific question, you may not see all of the conditions that apply and that are absolutely essential to understanding the phenomenon. The phenomenological perspective does not involve predetermined requirements for data acquisition and studying phenomena; it only requires that the method chosen should be relevant to the phenomena being studied.

One obvious choice, given phenomenology's inductive and ideographic perspective, would be to use ethnographic methods. These would help to avoid reducing material to a single theory, explanatory structure or area, and create the opportunity to see phenomena in all their complexity and multiple manifestations.

The truth of the phenomenological analysis is evaluated according to the extent to which the description of phenomena is coherent with the propositions that the researcher, through *epoché*, has collected about the phenomenon, and whether these propositions are sufficiently exhaustive and in-depth. The evaluation is therefore linked to whether the analysis's descriptions of the world/phenomena are consistent with what the people living in it describe, even if the descriptions in the scientific presentation make use of more abstract formulations than are common in colloquial language.

Phenomenology's credibility is also closely related to intersubjectivity, on two different levels:

- Firstly, on the level of analysis. Here, intersubjectivity is ensured through transparency in the research, i.e. it must be possible for researchers other than those involved to follow the researcher's analysis and check every detail.
- Secondly, on the material level, where a phenomenon can only be understood in depth through its many different forms of manifestation. Here, it is crucial not to reduce the phenomenon to the parts where the different forms of manifestation overlap, but to clarify the many different aspects of a phenomenon. Collating the various forms of manifestation therefore serves to illuminate the intersubjectivity.

Example of material collection and processing

In our example of changes in anti-poverty strategy in the World Bank and the famine in the South Sahara, we would choose methods of data collection and analysis strategies that were relevant to understanding the patterns of meaning in everyday practices, routines and habits of the three relevant groups – recipients, administrators and decision-makers – in connection with the change in poverty strategy before, during and after the famine. In a phenomenological perspective, all of the experiences and intentions are contextual and bound to a lifeworld, and therefore the data acquisition is directed towards encapsulating this lifeworld.

In this example, you could conduct interviews with appropriate representatives of the three groups and talk to them about their experiences, memories and perceptions of what happened, how they reacted to each other and their respective lifeworlds. In these interviews, it is very important not to add your own interpretations or ask questions that steer the interviewees' understanding of the process in a certain direction. The questions must be focused on the individuals' understanding and intentions. Semi-structured or fully open interviews should therefore be used in order to create *epoché*. The interpretations of events, and the concepts and words used to describe them, help us understand the interpretations and lifeworlds that the various groups used to help them understand the events.

At the same time, you could also make use of observations – e.g. of the Bank's council or of how decisions are made in the World Bank, the beneficiary organisations and NGOs – in order to arrive at a tangible understanding of practices and routines based on practice that is not revealed via interviews, but that helps establish the frameworks for day-to-day work.

Since the strategy change also involved financial transactions, it is relevant to describe the size and degree of financial aid as well. This could be done by document processing. It is crucial that the researcher does not automatically fit all of the data into a theory about what happened, but approaches the material impartially. The credibility of the analysis is ensured through the participation of the various groups involved in the phenomenon, and by letting their different voices be represented. Even though the analysis will be more abstractly expressed than the individual interpretations, it is important that those involved are not reduced to a single, collective voice. At the same time, you must also take care to describe not only what is used in the analysis but also what is omitted. This makes it possible to follow the different forms of reasoning in the analysis, and to evaluate whether or not its results are plausible.

CHAPTER 6

Structuralism and critical realism

Structuralism is multifaceted. Some forms of structuralism work with subject fields that are very close to positivism, in terms of their use of laws, while others are closer to phenomenology and hermeneutics. The Marxist-inspired version of structuralism is a critical perspective that seeks to counteract oppression and promote collective emancipation. The common denominator to be found in the many different strands of structuralism is an interest in the underlying structures that shape the way in which the world works. Structures can be linguistic, economic, social, psychological, etc. Structuralism has influenced linguistics, anthropology, the social sciences (particularly Marxism) and psychology (particularly psychoanalysis). In recent decades, *critical realism* has emerged as a separate field based on structuralism. Accordingly, structuralism may be wide-ranging; there are, however, many similarities.

1. In general, structuralism and critical realism are interested in identifying the structures that define the limitations and expressions of the objects studied. Structures exist in all natural science, social science and humanities contexts. At a very abstract level, structuralism adopts a new unity-of-science perspective and ontological reduction.
2. The subject field is a term that describes both the collective and the structures that form the basis for the collective's actions. The aim is to reveal something general about how

structures define people and their actions. Structuralism and critical realism are, therefore, *nomothetic*. Some of the perspectives consider structures to be objective, while others deal with how to change (unjust) structures.

3. The ontological basis for (most of) structuralism and critical realism is realistic. This means that structuralism assumes that the structures and interrelationships that it studies exist in reality, even if they cannot be observed directly. The aim is, therefore, to uncover the invisible – and often unconscious – structures and mechanisms that exist in the world.

4. Structuralism and critical realism aim to understand and explain at the same time. It would usually be argued that the structures we attempt to uncover exist independently of the researcher, even if some of them are man-made. Since the structures are invisible and unconscious, they cannot be known empirically, but must be divulged by identifying recurring patterns and elements. This kind of knowledge usually requires that we amass and process a very large amount of data from which conclusions can be drawn, both deductively – i.e. hypotheses about presumed structures are verified – and inductively, i.e. repeated patterns are explained by establishing theories about the structures that generate them.

5. The view of human nature that characterises structuralism and critical realism considers humankind as subject to structural conditions in which the individual forms part of the collective. As such, individuals are not looked at in their own right *per se*, but as a reflection of the community to which they belong. In this way, individual intentions are not viewed as individual, but as part of the logic of the (invisible and unconscious) structures. Certain theoretical currents within structuralism and, especially, critical realism are reappraising the relationship between structure and individual.

6. The truth theory that applies to structuralism and critical realism is a representative version of correspondence theory

– i.e. propositions are considered true if it is likely that the structures revealed actually do exist independently of the researcher's interpretation. However, these structures can never be identified empirically (i.e. we cannot observe the structure itself), and therefore the truth of the propositions can only be verified by interpreting the repeated patterns in the data as representations and reflections of structural conditions.

The roots and kindred spirits of structuralism

This chapter outlines structuralist thinking, starting with its various pioneers, and incorporates perspectives such as linguistics, anthropology, Marxism, critical theory and critical realism. These traditions have largely been based on specific disciplines, but they have also been very important to structuralism or structuralist understandings in other disciplines as well. They are divided into two perspectives: one sees the world as a set of symbols (linguistic structuralism and structuralist anthropology), while the other focuses on material structures (Marxism, critical theory and critical realism). At the end of the chapter, the main structuralist and critical-realist concepts are brought together and related to our specific case study.

Structuralism in linguistics

The Swiss linguist Ferdinand de Saussure (1857–1913) is often called the father of structuralism. His work *Cours de Linguistique Générale (Course in General Linguistics)* was published in 1916 by some of his students. Based on comprehensive lecture notes, it presents Saussure's radical approach to linguistics, which marked the start of structuralist thinking.

In Saussure's day, linguistics tried to explain all contemporary elements and changes by positioning them historically (*diachronically*) in relation to previous languages and words – or alternatively, explained changes with reference to psychological or physical resemblances between words and the world. Saussure's contribution is his separation of language into two distinct concepts. *Parole* is the actual spoken language, while *langue* consists of the underlying rules (grammar, phonetics and syntax) of the language system, of which we are not consciously aware but which mean we are able to talk to and understand each other. Rather than explain language historically, Saussure aims to explain its underlying rules systematically, through a simultaneous *(synchronous)* cross-section of how the language is used. In this way, he tries to reveal the invisible rules *(langue)* that govern the specific language use *(parole)*.

Saussure's description of language and words is based on language being a set of symbols. This means that words serve as symbols that point beyond themselves to something else. According to Saussure, the individual symbol consists of an expression *(signifiant)*, for example, the letters HORSE or word sounds [hÐ:s], and a content *(signifié)*, which is not the object itself, but the idea in the mind of the speaker and the recipient – in this case, the concept of a horse. Where previous research took for granted the idea that there is a relationship between the symbol used and that to which it specifically refers, Saussure identifies that the relationship between a symbol and what that symbol means is arbitrary (random). This means that, in principle, we could call the beast in the field something completely different. However, we cannot simply swap the symbol for any other one, because language is a social phenomenon that is determined by conventions. We cannot speak a language that only one person understands – at least, not if we want to communicate with others. That we understand what it means when we say HORSE is established through the system of differences that constitute the language system. HORSE is different from all the other symbols, and it is through differences with other symbols that each symbol derives meaning – HORSE assumes its meaning because it differs from all other signs, such as COW or CHILD.

Although Saussure's thinking is rooted in language, he developed an understanding of symbols that is rooted in systems of differences, which therefore extends beyond language to other sets of symbols. Several of his points concerning langue/parole, synchronic/diachronic and signifiant/singifié have been used to identify and analyse other sets of symbols.

Saussure uses chess to describe his understanding of language as a set of symbols. The chess pieces and the chessboard can be made of anything (wood, marble, paper, etc.), as long as there are the correct number of squares on the board and the correct pieces, each of which has its own meaning (pawn, king, etc.). If we want to understand chess, we have to know the invisible rules that govern the way we move the pieces. It is therefore not the actual game of chess *(parole)*, but the rules of the game *(langue)* that set the framework for how chess is played. In order to understand the rules, there is no point in studying the historical development of the game, that is, *diachronically*. Rather, we must study it on the basis of the pieces' mutual relations, i.e. *synchronously*. To understand the rules of the game, we must observe a large number of different specific games – which is, indeed, all we can observe – and, in the light of a reduction of these many different instances, identify the rules of the game. The individual chess pieces derive meaning through their relations to and differences from the other pieces in the game. It is not a law of nature that a pawn is a pawn. Rather, it is arbitrarily defined by convention that the pawn's significance is established through its relations to and differences from the other pieces.

Structuralist anthropology

In the 1940s, the French anthropologist Claude Lévi-Strauss (1908–2009) began transferring Saussure's ideas about symbols to anthropology. He took the large quantity of readily available information about family relations in pre-modern societies and reduces it to a few kinships that, in his opinion, regulate and organise society. Lévi-Strauss pointed out that all of culture's kinships or social organisations are structured around a system of different universal opposites – i.e. male/female, child/adult, permitted/prohibited, human/animal. This means that, although we are dealing with different cultures, whose differences are signified in a range of ways, they all relate to and organise around universal (binary) pairs of opposites. The importance of the elements is established by the contradiction within the binary opposites (*allowed* assumes meaning relative to *prohibited*, and *man* in relation to *woman*) and through their relation to other pairs of opposites (in other words, life/death is related to allowed/prohibited). Therefore, real changes in cultures will always be abrupt, because if one of these fundamental relationships is altered, then society's whole form of organisation changes too.

This reduction of human life to a few basic pairs of opposites means that what might on the surface appear to be a change, in reality is not, because it has not changed the basic relationship. This means that in Lévi-Strauss's universe, change and development cannot take place at the same time but occur due to abrupt breaks between eras that are radically different in their form of organisation. As such, current historical developments are deemed to be merely ripples on the surface – there is no significant challenge to the fixed structures that constitute the range of possibilities for the ideas expressed or actions performed.

Lévi-Strauss continued this universalist thinking in *La Pensee sauvage (The Savage Mind*, 1967), in which he argues that the cultures we call irrational or primitive merely express a different logic and rationality, which is organised differently than the one with which we are familiar. The anthropologist's task is to uncover the rationale behind the apparently incomprehensible and irrational actions and thoughts that we observe. This universalism in both Saussure and Lévi-Strauss is a reaction to the hitherto dominant historical explanatory framework in their discipline, which was based partly in hermeneutics.

Marxism's basic concepts, and its offshoots

Although Marxism was established before Saussure's structuralism, it contains several elements that belong to the same perspective, in particular, the division of reality into two distinct parts: empirical manifestations and underlying structures. Like structuralism, Marxism distinguishes all of the phenomena that we can see and observe – i.e. those dealt with by positivism – from the essential, underlying and governing structures that constitute the real object of science. Between the 1960s and 1980s, Marxism and structuralism enjoyed a resurgence in university courses and played an important major role in all academic disciplines at the time.

Marxism stems from Karl Marx (1818–1883), who is particularly famous for texts like *The Communist Manifesto* of 1848, written with Friedrich Engels, *Capital (1867, 1885, 1892)* and *Critique of Political Economy* (written in 1857–1858). Marx's main thesis, which formed the basis of Marxist thought, is that all social formations are composed, on the one hand, of a *superstructure* (consisting of ideology, law, politics and forms of consciousness) and, on the other, of a material basis that forms the basic economic structure. This basis reflects the relationship between two universal categories: (1) forces of production (raw materials, labour, energy, machinery and means of production), and (2) relations of production (organisation of productive forces, ownership). The relationship between the two will change regularly and lead to social revolutions and new relationships between the forces and relations of production. It is the material basis that determines how the superstructure looks. Marxism is therefore based on a strongly materialistic understanding of science.

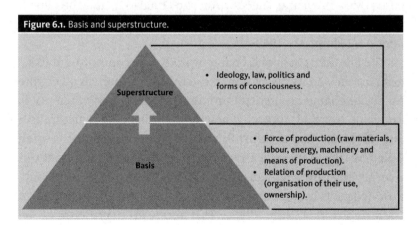

Figure 6.1. Basis and superstructure.

The basic features of Marxism include the concepts that all facts are social facts, and that the collective and society form the basis of scientific analysis, and therefore must be studied as such. Marx's theses include a form of causality that aims to predict and generalise all social formations through time and across space. This causal understanding was later abandoned by neo-Marxism.

Marxism places particular emphasis on scientific knowledge's ability to contribute to social change. In Marxism, research has an objective, namely making the workers – or, to use Marx's own term, the proletariat – aware that their lives would be different were it not for the unfair distribution of goods in society, which is only maintained due to a lack of knowledge about the specific, oppressive relationship between employer and employee. This commitment to social questions makes political action a normative concept in research.

The "Frankfurt School" expanded upon Marx's thinking in a significant manner. It linked neo-Marxism with Freud's psychoanalysis, and was a major reason for the resurrection of Marxism and its critique of capitalist society from the 1960s onwards. The Frankfurt School comprises philosophers and sociologists at the Institute for Social Research in Frankfurt. It was established when the sociologist Max Horkheimer (1895–1973) was appointed head of the Institute in 1931. A differentiation is often made between the early Frankfurt School, including Horkheimer, Herbert Marcuse (1898–1979) and Theodor Adorno (1903–69), and the newer/younger Frankfurt School under Jürgen Habermas (1929–) and later Axel Honneth (1949–).

The Frankfurt School is characterised by its commitment to social critique and its discussion of the role of research as an engine for social change. Its central premises are that, firstly, science is not value-free and, secondly, nor should it be. Like structuralism and Marxism, the Frankfurt School (or *critical theory*, as it is often called) works to uncover the underlying logic and contradictions that govern capitalist consumer society. Its aim is to contribute knowledge to human emancipation from ideological structures – referred to as *ideology critique*. This also leads to criticism of the ideologies that control research itself. The Frankfurt School sees it as its mission to uncover (and ultimately improve) these ideologies through critical analysis of the world. In the world of the sciences, positivism has been a particular object of the Frankfurt School's critique.

Critical realism

Critical realism, a later structuralist perspective, arose in the 1970s as a critique of both positivism and hermeneutics. Subsequently, critical realism developed further as a critical corrective to social constructivism, which was making great strides (for more on this, see Chapter 8). Roy Bhaskar (1978, 1989, 1998) is considered the author of this approach, alongside Margaret Archer (1995), Rom Harré (1986), Tony Lawson (1997) and Andrew Sayer (1984, 2000). Critical realism confronts the unity-of-science concept on which the positivist and empiricist approaches are based. Like structuralism, critical realism claims that the object of all scientific fields is divided into surface and depth, and that they should also be studied in that way. However, it is important to start with what characterizes the specific object of study, and therefore this perspective is characterised by methodological pluralism.

Like structuralism and Marxism, critical realism takes as its starting point the idea that the world consists of several levels. Specifically, Bhaskar describes three levels. The first two levels together form the *intransitive* dimension, i.e. the ontological part – the world as it actually is, when it is not analyzed or discussed.

1. The *real* level – the deep structures and mechanisms that create the possibility of actions and phenomena.
2. The *actual* level – the events and phenomena that have been created on the basis of the structures and mechanisms at the real level. The relationship between the real and actual level is not a causal relationship that can be mapped in positivist laws. Like structuralism, it is a relationship that is open – i.e. multi-causal and contextual – and it can there-

fore be described and explained, but it cannot be used to predict future incidents, events and phenomena.

3. The *transitive* part/empirical level. This is made up of the empirical level, which describes the phenomena and events that we experience and are able to observe. This level is the epistemological part. It is here that we develop theories and scientific analyses about the world that are inevitably coloured by our beliefs and perceptions. This does not mean that there is not a real world (no. 1 and no. 2) that we must try to explain and describe. Rather, it means that we can never be sure that our scientific findings are true and in accordance with reality, even although they are included in the reality that they describe. As Bhaskar points out: "Social science does not create the totalities it reveals, although it may itself be an aspect of them" (Bhaskar 1998: 47).

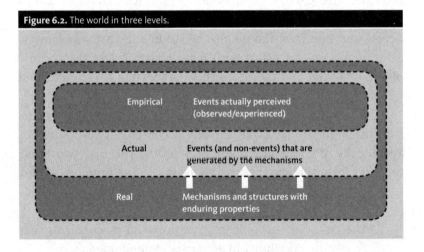

Figure 6.2. The world in three levels.

This view of the relationship between depth and surface also affects how critical realism looks at the relationship between structure and individual. On the one hand, there are structures and mechanisms that set the range of possibilities for the individual's actions, but on the other hand, the individual constantly interprets those structures and mechanisms through his or her actions, and in doing so potentially helps to change them. There is, therefore, a kind of interaction between structure and individual.

The central paradox here is that knowledge is dependent on the people who create it (as discussed previously, science sets parameters but is also constantly (re-)defined through tangible actions), but the actual object of scientific investigation is independent of the knowledge production. As such, we are dealing with a combination of ontological realism (things exist independently of the observer) and epistemological subjectivity (the actual observation is subjective). This has implications for how we acquire knowledge in a critical-realistic perspective.

Where we have previously talked about deduction and induction, critical realism speaks of *retroduction*, which can be equated with Peirce's abduction (more on this in Chapter 9). In short, rather than distinguishing between deduction and induction, the movement from the abstract/theoretical to the specific (deduction) and from the specific to the abstract/theoretical (induction) is seen as a combined effect necessary in the research process. Since the deep level cannot be directly observed, we must acquire access to it in some other way: (1) by assuming that the deep level exists, we are therefore assuming that there is an underlying structure to be examined; (2) establish hypotheses on the basis of other contexts, similar to the context we want to study; and then (3) test the hypotheses' usability against the empirical data that it is possible to collate. If the hypotheses prove to be sustainable, this can be seen as an indication that the underlying structure – the basis for the hypotheses – exists.

Critical realism draws on both Marxism's and critical theory's requirements that science must relate critically to society and the world in which we live, and that knowledge of the world is key to understanding and challenging unfair conditions. Since, according to this perspective, we cannot predict anything, science is not a template for action – but it does identify the structures that oppress people.

Establishment of the subject field in a structuralist and critical-realistic perspective

In both the structuralist and critical-realist perspectives, the subject field consists of the collective and the structures and mechanisms that form the background and provide the opportunity for collective actions. Both structuralism and critical realism challenge positivist unity-of-science thinking, which is based on empiricism. Instead, a new understanding of unity of science is introduced, in which all phenomena must be understood at a deep/surface level, and must be studied differently due to the range of subject areas that the disciplines work with in practice. The aim is to generalise about how structures form phenomena, people and their actions. Therefore, both structuralism and critical realism are highly *reductionist* and *nomothetic*. It also means that structuralism and critical realism are highly anti-phenomenalist – i.e. we are not interested in how phenomena are perceived by the people who experience them, but in what they really are once we peel away their surface.

The ontological basis for (most of) structuralism and critical realism is, as mentioned in the introduction, realistic. This means that structuralism and critical realism basically assume that the structures and relationships they study exist in reality – even if they cannot be observed directly. The purpose is to uncover the invisible and often unconscious structures and elements that exist independently of the researcher. Marxism and critical realism, in particular, consider the structures to be material, whereas structuralism incorporates both materialist and idealist positions.

The view of human nature in structuralism and critical realism is that humans are subject to structural conditions in which the individual forms part of the collective. Individual intentions are not individual, but shaped by the inner logic of the structures. The individual barely sees or is aware of the inner logic. Several theoretical currents within structuralism and critical realism are working to rethink the relationship between structure and individual. The aim is that this relationship should not be deterministic but open and flexible, a relationship in which the individual is contextualised as a user of the structures s/he both is shaped by and continuously forms through his/her use of them.

Example of a research question

Using the World Bank's anti-poverty work as a basis, and working from a critical-realist and structuralist perspective, you can pose the following research questions:

- What power relationships are established between the World Bank's donor countries and recipient countries?
- What is the current cultural understanding of the relationship between recipient and donor countries in the World Bank and its anti-poverty work?

In the critical-realistic research question about the balance of power, you would usually look at the material conditions that form the basis for any domination and dependency interrelationship. In this context, the balance of power could be seen in the light of the relationship between the different countries' natural resources, production and trade, as well as the World Bank's political and economic initiatives. Your aim in this case would be to identify how the various measures reflect underlying power relations that are not directly visible. Critical realism – because it considers structures to be open and flexible – also focuses on the relationship between structure and actor. In this case, your interest will be in both identifying the underlying structures that govern the power relationship, and how the recipient countries' reactions and use of money may help to transform this relationship. More generally, the study could identify how power structures produce the scope for action within which the recipient and donor countries are able to act. Here, it is not certain that the change that appears to take place on the surface level also signifies a deeper change. The aim is to render visible those invisible factors that regulate and produce relations in the international community

– both economic and political – and thereby to introduce the possibility that they can be changed.

In the structuralist research question, the subject of study is the unconscious and cultural perception of the relationship between recipient and donor countries. Your focus will be on how the various discussions about payments and actions reveal the underlying structures that help to create the range of possible actions within which people think and behave as they do. You are therefore interested in uncovering the rationales that lie behind the actions and the discussions of, and arguments for, these actions. Basically, the idea that something is legitimate, natural or rational is based on a cultural understanding of the world that is reproduced through, for example, the World Bank's anti-poverty initiatives.

Knowledge generation in a structuralist and critical-realist study

Structuralism and critical realism seek to understand and explain at the same time. It would usually be argued that the structures you are attempting to uncover exist independently of the researcher, even if some of these structures, e.g. scientific knowledge, are man-made. Therefore, you will methodically separate the researcher from that which is being studied. Since both structuralism and critical realism are concerned with invisible and often unconscious structures, their subject fields cannot be recognised directly and empirically, but must be discovered through recurring patterns and elements, the existence of which can be explained with the help of the invisible structures.

Usually, this kind of knowledge requires the processing of very large amounts of data at surface level (the only level to which you have access), which is used to access the deeper structures that you want to uncover. Conclusions are therefore reached both deductively – by verifying hypotheses about structures – and inductively – by processing large amounts of data to explain repeated patterns through theories about the structures that generate them. The strength of the knowledge generated is evaluated according to whether it can reveal invisible and oppressive structures. Critical realism combines the deductive and inductive approaches, and is connected – via retroduction – to analogue hypothesis formation (i.e. hypotheses formed in relation to other similar contexts).

Large parts of structuralism and critical realism – notably, of course, Marxist structuralism – maintain that science can never be value-free. In this perspective, the role of science is to generate knowledge about the oppressive structures under which people live, thereby creating a breeding ground for emancipation and social change.

Structuralism and critical realism account for the plausibility of their conclusions through a kind of correspondence theory, which, in principle, evaluates the conclusions on the basis of how well they match the empirical data – you cannot see the structures, but you can evaluate how well the hypotheses about the structures accord with the empirical data. Statements are therefore considered true if they make it probable that the structures that have been identified actually exist independently of the researcher's interpretation. However, as these structures can never be identified empirically, the truth of the propositions can only be verified by interpreting the patterns that recur in the material studied as an expression of structural conditions.

Example of data acquisition and processing

In the example of the power relationship between recipient and donor countries, you could try to uncover the material conditions that maintain the power structures between recipients and donors. This could be done by looking at different resources (raw materials, level of education, institutions, infrastructure, etc.) and the ownership of them in each country – e.g. are they private or state-owned? You could collect data on these issues in both recipient and donor countries and, in this light, try to identify patterns that indicate the power relationship, based on material conditions. For example, you might ask: what power relations exist in the resource relationship between recipient and donor country? What role does the donor country play, e.g. in relation to resource extraction in the recipient country? How does education contribute to this process? You could then look at whether this relationship is maintained through donations. To whom do they donate and why? Is the aim to improve the level of education, or is it to exploit the country's resources? Which type of ownership is strengthened in this way? What requirements are placed on receipt of financial aid? Does the aid serve to change or maintain the power relations? Such a study will be based on certain power relations existing, but their actual content not being known. You could generate hypotheses based on analogies from other similar studies, and then test the hypotheses against the material you have collected. In this context, the material will be quantitative, measurable data that can be analysed for any (inter) relationships. The credibility of the analysis is tested in relation to whether your hypothesis is accepted or rejected on the basis of the empirical material.

The example of the cultural understanding of the relationship between recipient and donor countries calls for a different kind of material. In this case, the relevant data might be descriptions – by the World Bank's staff and donor governments – of the anti-poverty work, especially the way in which it is legitimised, for example, interviews, legislative proposals and procedures. In this case it is the idealistic level that must be identified. In a structuralist perspective, culture is unconscious, and therefore employees cannot be asked directly which culture they perceive as crucial in the fight against poverty. However, the answers to such questions will be regarded as a reflection of underlying understandings that trigger the actors' actions and descriptions. Again, there is a need for a certain amount of data in order to draw more general conclusions. You might, therefore, want to extend the analysis with a survey and include questionnaires that aim to reveal the extent to which the culture in question is common to different groups.

CHAPTER 7

Realism and constructivism

The philosophy of science perspectives with which we have worked so far maintain that the world and the various phenomena studied have some kind of essence. This essence may be idealistic (i.e. to do with thoughts and ideas), materialistic (i.e. physical and sensory) or both at the same time. In positivism and critical rationalism, causal connections do actually exist in the world. For hermeneutics, human cultural understanding of the world is an essence, which we as researchers seek to uncover through interpretation. Phenomenology seeks to understand the experiences from which phenomena emerge, and understands essence as existence. Structuralism and critical realism believe that the structures and mechanisms they seek to reveal exist and affect how people act. The -isms covered in the chapters that follow highlight the problems associated with the concept of essence, and some of them question whether it makes sense to distinguish between essence and non-essence.

This chapter focuses on the critique of the realist position inherent in the "linguistic and cultural turn" that gained ground and had wide-ranging consequences in the humanities and social sciences from the late 1960s onwards. The critique revolves firstly around *epistemological* questions that have led to various forms of relativism – i.e. questions about whether it is possible to talk about and explore the world as an independent entity that exists outside of our knowledge of it. Secondly, the critique is related to the fundamental *ontological* question – namely, whether we assume that the things we study have essence per se, which we as researchers must attempt to understand/identify/know, or whether they can be said not to have an essence per se but instead represent or reflect our categorisations, definitions, classifications and interpretations of the world.

The principles of realism

Realism's starting point is that the world exists independently of our consciousness and ability to describe it. As a result, our hypotheses about the world should also be tested for their veracity – in other words, it should be possible to either prove or disprove whether what we say about the world is true or not. In realism, scientific truth is considered to be universal. This means that statements about the world are true or false, irrespective of where and on what basis we observe the world. A statement about a global economic crisis made somewhere in the world in the period 2008–2012 cannot be true in one location or in one theory, and false somewhere else or in another theory – even though different theories propose different ideas about what is true and false. Whether a statement is *actually* true or not must be determined by testing it in the world, on those who propose it or via proxies. It is also the case that something can be true even though no one has observed it or figured it out yet. The world and its events and phenomena exist in themselves. The researcher's task, therefore, is to use diverse and rigorous methodological approaches to understand and/or explain them as accurately and clearly as possible.

Within the different versions of realism, opinions vary about whether the subject can ever gain access to the truth about reality. *Naïve realism*, which assumes that it is possible to know the world as it is through the senses alone, is often mistakenly ascribed to positivism, and is subsequently opposed. In reality, few philosophers or scientists believe we can know everything about the world directly via the senses alone. However, much positivist-inspired research is based on scientific and representative realism, in which scientific concepts and interpretations are used to identify objective phenomena and contexts that exist regardless of whether we are aware of them or not. Hence, realism argues that scientific concepts and descriptions of the world are based on reality. For example, tuberculosis and ADHD are conditions we have discovered, in the same way as we have discovered the Middle Ages as a particular historical epoch with particular features, or the neuron as a component of the physical world. Or, in our case study, that poverty exists and can be identified.

Critics of realism point out that doubts about our ability to naïvely and realistically represent the world as it is imply that we must conclude that it is impossible to postulate anything true about the essence of the world. Based on this idea we cannot know whether our descriptions of the world are true, and we therefore cannot describe reality as it is, it is argued that the scientific designations and descriptions do not describe the world's essence. Rather, they must be considered relativistic – i.e. only true in certain contexts and on the basis of certain rationales, but untrue in other contexts – or the world's essence must be created through our description of it. Before I look at these criticisms in greater detail, I will briefly outline some of the ways of thinking which, qua their radicalisation, marked the shift from a realistic perspective to relativism and constructivism.

Historical background

The basic constructivist and relativist approach is based on Immanuel Kant's distinction between *das Ding an sich* and *das Ding für uns*. Kant identified the need to clearly distinguish what things are in themselves from what we as humans are able to understand and explain. The main point of Kant's distinction was to describe the basis for and limits to human knowledge, and his thinking has exerted huge influence in the humanities and social sciences.

The version of relativism and constructivism that emerged from the cultural and linguistic turn took the next step and radicalised these questions. From the 1970s onwards, the exponents of what has been called the linguistic or cultural turn raised relevant fundamental questions. These were not new questions, but they were given renewed vigour by hermeneutics, phenomenology, structuralism and Marxism (Jantzen 1996; Gergen 1994: 30–64). As previously mentioned, all of these positions have a realistic foundation, but much of their thinking had been steered in a more constructivist direction.

The cultural turn placed cultural phenomena at the centre of the agenda, and introduced anthropological methodology and insights into many disciplines. Due to hermeneutical and phenomenological statements that human actions are based on people's experiences of phenomena and the meanings they ascribe to various events, and due to structuralist anthropology's focus on culture as a vital structural framework, culture was employed in research as both an explanatory framework and an analytical tool. It was used to uncover how and why people, at different times and in different contexts, understood events and phenomena differently from the researcher and from other people in other places and at other times. This led not only to other cultures' understandings of topics such as the body, gender, death, life and knowledge becoming the subject of historical and anthropological narratives about other places and times, but also to a more general idea that our basic assumptions about human existence are not innate and natural. A body is not something inherently natural that is merely perceived differently at different times – rather, it is the very essence of the body that differs and, as such, is thought about, experienced and treated differently. In other words, there is no essence beyond knowledge.

The cultural turn and relativism

Awareness of cultural issues was therefore not just another part of the scientific repertoire, on a par with all the other things that can be studied. It attained a privileged position because all phenomena and all contexts could and should be studied on the basis of the cultural rationale of the people being studied, i.e. their understanding and description of everything from economic crises and art to weddings, meals and agriculture. The cultural turn formed fertile ground for a relativism based on different rationales, different knowledge and different understandings of the world, which implied that things could be true for some people and untrue for others. In the radical version of relativism, it was impossible to evaluate which knowledge was true, since no knowledge is based on universally essential conditions, and therefore all knowledge is, in principle, only true from a certain perspective.

In the beginning, relativistic knowledge was only used to identify the culturally embedded rationales, knowledge and actions of the people and groups being studied. Later, however, the question was raised of whether the researchers' scientific knowledge should also be studied on the basis of the same cultural awareness. Research had to be deprived of its privileged position and considered a reflection of the cultural world and the rationale in which it was embedded. Research could not provide a definitive answer about what and how the world really was, and relativism was no longer simply a background perspective for recognising the phenomena studied and the contextuality of events – it also found its way into the researchers' practice. As a consequence of the radical version of relativism, the question was asked whether scientific knowledge can be raised above other knowledge, and how we could guarantee that the researchers' truths were not only valid in their own time and in relation to a particular rationale. As such, there was a shift from local and object-focused relativism – in other words, that truths are rooted in different cultural rationales, different times and different places – to a universal relativism, according to which no truths are more correct than others, because all truths are inevitably embedded in their historical and cultural context.

Inspired by structuralism, the focus of study also shifted from the individual to the individual-in-society. Individuals could not, as in the hermeneutic and phenomenological approaches, be understood by virtue of their intentions. Rather, they had to be understood through the collectives they entered into and were a part of. The 1970s relativism was also characterised by a socially critical perspective based on Marxism, critical theory (the Frankfurt School) and critical realism, which studied how ideologies and power structures not only shape our actions but, on a more fundamental level, determine peoples' lives, tastes and understandings (Gergen 1994: 44ff). In this way, relativism underpinned critical research that aimed to uncover how our society is not natural and inherent, but the outcome of ideology and power relationships. Where Marxism and structuralism partly placed the researcher outside the field of study and provided an opportunity to study how ideologies and power relationships govern people's actions, knowledge and scientific production, post-structuralism argued that researchers are never able to liberate themselves from power relationships (Gergen 1994: 53ff). As a result, the post-structuralists pointed out that the researcher exists in a web of predefined ideologies, structures and mechanisms, and therefore the critical research strategy must also address scientific knowledge itself.

The above description is a little blunt. There are many different positions, and many of the researchers labelled post-structuralists are less clearly relativistic. However, the key point is that the relativism that emerged in the early 1970s gave rise to a view of science that is sceptical not only about the validity of scientific propositions (e.g. phenomenology's scepticism vis-à-vis positivism), but about the possibility of objectivity in general.

The linguistic turn

The linguistic turn cannot be clearly distinguished from the above movements, but is based on an idea from Ferdinand de Saussure's system of signs (described in the previous chapter), which rejects the idea that language acts as a window through which we see and describe the essence of the world. According to Saussure, language's representation of reality is arbitrary, since the relationship between the world and language is established through conventions. Words do not derive their meaning and significance from agreement with that which they describe, but through their relation to other words in a system.

In the 1960s and 1970s, studies were often based on Saussure's concept of signs, and analysed many other aspects of the world as sign systems that could be read as if they were a text. The expanded text concept was both a theoretical and methodological technique that considered wars, revolutions, wills, social institutions and the Counter-Reformation as texts, i.e. as conditions that followed a certain "grammar" *(langue)*, which could be identified (through *parole*) on an abstract level. The corollary of Saussure's thinking is that language and reports of events could not be considered to be representations of what had actually taken place, but rather representations of the conventional way certain people understood, organised and attributed meaning to events that take place in the world. This led to the use of language and narratives about events being used as analytical tools with which to gain access to other people's rationales, etc. These ideas changed the relationship between language and the world. Previously, it had been argued that scientific propositions represented the world that they described, whereas Saussure's system of signs asserted that this link was arbitrary and only established by convention – in other words, it was merely indicative of consensus.

This was important for how to understand the relationship between what was said and what it was said about – and consequently, whether it was possible to speak about the truth and essence in the world's phenomena (Gergen 1994: 48–52). Language was no longer considered a window on the world but a reflection of the point from which it was observed (Jenkins 1991: 33–70). The status of language changed, from providing access to understanding the world to preventing this understanding, because we could not talk about anything outside language. This raises the question of how we can claim that what we say about the world is true and a reflection of what exists in the world if it does not correspond to the world and derive its meaning from it. Saussure's philosophy of language played an important role in thinking about truth and truth concepts.

Critique of realism

The questioning by the cultural and linguistic turn of whether researchers are able to attain true knowledge about the essence of the world has had a variety of consequences, for instance, anti-realism, which takes the same starting point as realism, namely that the truth of a statement must be tested in reality and that truth must be established universally through correspondence with the world. However, unlike realism, anti-realism concludes that we cannot test the truth of our statements on the world and thereby identify its true essence. This means that there is no universal truth, essence or certain knowledge – and we must consider and study how all knowledge (including our own) rests on certain power relationships that favour certain forms of knowledge at the expense of others. As we will see in Chapter 8, these are the kind of considerations that a structuralist and a social constructivist take into account.

More moderate critics of realism spend less time on the question of truth and relativism. One example is *pragmatism*, which originates from 19th-century American philosophers such as Charles Sanders Peirce, William James and John Dewey (see Chapter 9). The basic idea here is that it is not so important whether our knowledge is of general application. The key is what effect it has in practice, and what science postulates about reality is seen as a series of practical thinking tools. In its more moderate form, this means that scientific hypotheses are not necessarily true, and that we cannot determine whether they are. We therefore need to evaluate scientific concepts and hypotheses on the basis of whether they are usable and useful (this kind of pragmatism is often associated with the American philosopher Richard Rorty), or whether they can be made more probable through coherence with the connections in relation to which we describe them (with reference to the different truth theories introduced in Chapter 2). Newer and more radical approaches (including STS and ANT, see Chapter 10) attempt to establish brand-new definitions and intersections between what is considered to be real and what is regarded as constructed, and whether one excludes the other.

Different degrees of relativism

To recap a little: relativism criticises scientific realism's universality. It claims that there are no universal truths about the world's essence and rationales, since humankind is the measure of all things (as the Greek Sophists were accused of proclaiming). Often, relativism is based on the fact that there are many and inconsistent rationales and ways of living, and is thereby very close to cultural relativism, which means that we can only understand a group, a nation or an individual on the basis of the norms and rationales by which that group or person personally live(s) – we cannot evaluate or understand them in relation to a rationale or a truth that can be said to apply to all people at all times. This means that we cannot assess whether poverty exists in a country on the basis of universal considerations. It depends on what rationale people adopt in particular contexts and at particular times when talking about poverty. In their radicalism, these ideas may lead to the dissolution of the reality concept and to a discussion of how realities vary across different spaces.

The underlying assumptions of the relativistic perspective imply a fragmentation into various (almost) closed groups of particularities, each of which can insist on its own reality and truth. We cannot judge whether one group is more right than the other – both may be right, but within different rationales. As such, this represents an epistemological critique of the universal truth concept.

There are different degrees of relativism. Local relativism argues that relativism only applies within the domains of human life that have to do with social contexts, and disregards physical contexts. Global relativism, on the other hand, is based on the assertion that everything is relative and depends on which rationale forms the basis for understanding the world. A typical – and obviously relevant – criticism of global relativism is to subject it to itself and say that relativism is also relative.

Constructivism

In Chapter 8 of this book we will look at the social constructivism tradition and its influence in greater detail. However, at this juncture it worth casting a quick eye over constructivism in a broader sense, in the light of the critique levelled at the realistic perspective. Constructivism's critique of scientific realism is rooted in scepticism about whether scientific statements can be valid independently of how people make sense of the world. The starting point is what is known as the cookie-cutter theory, according to which the world is basically a formless dough that we shape via our language and analytical concepts (e.g. "society" or "nature"). We shape parts of this dough into pastries but leave other parts unused. In other words, we define some of the shapeless mass as phenomena that we can study, learn about and act in relation to, whereas other parts of the dough cannot be described and therefore cannot be understood and dealt with. Knowledge, then, is an active and constructive action. Or, to use a well-known phrase: "Knowledge is power" (Foucault 2002 – see Chapter 8).

Constructivism means many different things and ranges along a continuum on which one pole asserts that the significance people place on events affects their actions, and another, more radical pole invokes an ontological critique. One famous example of the ontological critique is by the French philosopher Bruno Latour: We cannot say that the Pharaoh Ramses II died of tuberculosis because the tuberculosis bacterium was not discovered by Robert Koch until 1882, nor does it make any sense to claim that diseases in 1230 were not caused by an imbalance in the four cardinal fluids and cured by restoring equilibrium, e.g. by draining the patient of blood or bile – which was precisely what people believed at the time. More generally, constructivism implies that knowledge is not generated by the essence of reality, but that statements and concepts about the world are legitimised by social processes and therefore created collectively.

Different traditions of constructivism

We can divide the different constructivist traditions into four main categories:

- Discursive/communicative construction
  - ◊ In this category, realities are considered to be constructed through language, narration, metaphor, concepts and rhetoric. Prominent exponents are Michel Foucault, Kenneth Gergen and discursive psychology. Here, it is the concepts that form the dough (the world) into specific shapes (this approach will be explained in greater depth in Chapter 8).
- Social construction
  - ◊ In this category, the starting point is that it is social groups that create realities collectively, and that these realities are shaped by the groups' interests and goals. Special attention is also therefore paid to the power that individual groups accrue via their constructions. For example, this can be the power that privileged groups acquire by creating a certain understanding of democratic institutions that benefits them. This power always meets resistance, which leads to conflict regarding which groups define how we understand various phenomena (e.g. democratic bodies), and on the basis of which interests they do so. Here, it is a group's interests and objectives that shape the dough. Important exponents of this approach are Peter Berger, Thomas Luckmann, David Bloor, Steven Shapin, Simon Schaffer, and Harry Collins (this approach will be explained in greater depth in Chapter 8).

- Cognitive construction
  - ◊ This approach assumes that our consciousness or neurological development makes certain kinds of worlds accessible to us. With the famous statement "What the frog's eye tells the frog's brain", J.Y. Lettvin et al (1959) pointed out that the frog's interest in protecting its life has significance for what it sees and how it perceives the world. This way of thinking forms the background for an evaluation of people's perception of the world. Here, the cognitive structures, be they biological or conceptual, shape the dough (the world). For example, on the basis of experimental studies of children's developmental stages, the Swiss psychologist Jean Piaget (1896–1980) developed a theory that people store knowledge and understandings of the world in various types of consciousness schemata that have consequences for our actions. The work of the linguists George Lakoff and Mark Johnson (including the classic *Metaphors We Live By* from 1980) is based on the idea that the metaphors through which we understand our world are not just ways of expressing ourselves, but firstly stem from our basic perception of being a body in the world, and secondly are used to make sense of the world – e.g. when game or war metaphors affect our understanding of the phenomena. Important exponents are Humberto Maturana, Francisco Varela, Jean Piaget, Ernst von Glasersfeld, George Lakoff and Mark Johnson.
- Socio-technical construction
  - ◊ Socio-technical construction is based on the idea that the dough is shaped by the network that, through dynamic interaction between humans and non-humans (for example, technology), interweaves and connects different interests, ideas and actions to form a phenomenon. This includes science-technology studies (STS) and actor-network theory (ANT). A key element in this version of radical constructivism is that the more these domains

are constructed, the more real they become. Or, to use a better-known formulation: "The more constructed, the more real" (Latour 1987). Important exponents are Donna Haraway, Bruno Latour, Andrew Pickering, John Law, Michel Callon and Steve Brown (this approach will be explained in greater depth in Chapter 10).

It is important to understand that the discussion about realism is not new as such. In their more moderate versions, many relativistic and constructivist understandings form the basis for a large number of the traditions that I have previously called realist, and it was the continuation and radicalisation of their insights that led to the cultural and linguistic turns. Constructivism and relativism have had major consequences for the concept of truth in research and for discussions about the role of research in society. The focal point is, firstly, the question of how we relate to essence and non-essence; secondly, whether we can divide knowledge into different spheres and distinguish between natural, social and human phenomena; and thirdly, whether we can distinguish between ordinary human knowledge and the researcher's knowledge.

The next three chapters present constructivist and pragmatic traditions, and look at the different ways in which they deal with the question of essence and realism versus relativism.

CHAPTER 8

Social constructivism and discourse analysis

In recent decades, social constructivism has swept the academic world, making inroads into almost all social science and humanities disciplines to some extent or other. While social constructivism started off having a rebellious and liberating effect, questioning everything previously taken for granted and considered almost natural, today it has itself become a standard theoretical perspective in many disciplines.

Social constructivism and discourse analysis have many facets and feature several schools of thought that develop in different directions and combine relativism and constructivism in different ways. However, some common denominators are identifiable. Generally speaking, social constructivism and discourse analysis embody the following elements:

1. Social constructivism and discourse analysis reject the idea that some types of knowledge are more privileged than others, which renders the *unity of science* project irrelevant. Science is just one system of knowledge with its own associated truth markers, in the same way as journalism, religion and other knowledge systems.

2. The knowledge-constitutive interest lies in identifying the way in which thought and speech – as well as the everyday truisms people live with and under – are established, used and changed, with the focus on processes rather than enti-

ty. Social constructivism is characterised by its ambition to be *critical and liberating* by revealing the man-made elements of the truisms, thus facilitating change to undesirable conditions that may otherwise appear immutable.

3. The ontological basis for social constructivism and discourse analysis is constructivist. The key is how a phenomenon is created in a specific context – not the essence of the phenomenon per se. There are different degrees of radicalism. Some deny all forms of "natural" existence, while others are more open on this point. However, a common feature is attention to the way phenomena appear and the impact they have within their contexts, rather than their natural way of being.

4. The epistemology of social constructivism and discourse analysis is based on the idea that knowledge is always coloured by time and place, and therefore changeable. As a consequence, knowledge is considered both a result of the context and a tool with which to transform it. This implies a relativism that is present in different degrees. Instead of understanding what significance phenomena have, the idea is to understand how meaning is ascribed to phenomena, as well as explain how this process works. The studies are often inductive, as they take the specific context as a point of departure to define relationships, rather than examining pre-existing hypotheses about those relationships. Nevertheless, social constructivism is based on the theoretical assumption that the social context is crucial.

5. As a result of the focus on the role played by language, discourse and meaning in social constructivism and discourse analysis, science is not considered value-free. For some, this means that science – which necessarily defines a specific interpretation – must strive to create the basis for a better society that liberates people and provides the oppressed with a voice.

6. The view of human beings is characterised by the idea that human nature is not central. The individual is considered

as a representative of a general collective, defined by discourse and social constructions.

7. The truth theory that applies to social constructivism and discourse analysis is coherence theory, according to which a statement is true if it forms part of a system of interpretative statements (i.e. like hermeneutics) in a way that is free from contradictions. As a result, conclusions are drawn based on whether they are consistent with the overall theoretical framework and the concrete emperical material. It is the relationship between the different parts of the social constructivist and discourse-analytical analysis that ensures its credibility and reliability.

Before proceeding, a conceptual clarification is called for. In most disciplines, the concepts of social constructivism and social constructionism are used interchangeably. This is not the case in psychology. In psychology, constructivism denotes cognitive developmental psychology in the tradition of Jean Piaget (1896–1980) and others, and refers to the social interaction that takes place when individuals develop cognitively, learn or understand in interaction with other individuals. Due to this, in 1985, Kenneth Gergen suggested that the field of psychology should use the concept of social constructionism instead of social constructivism, when talking about what we have so far referred to as social constructivism. This book uses the term social constructivism in the sociological, not psychological sense, unless otherwise noted.

The degree of constructivism

It is important to stress, that social constructivism and discourse analysis are not schools with a shared source of inspiration, but a multitude of different approaches that seek to explore how reality is constructed socially and discursively. Three different axes are identifiable, however:

- The first axis deals with the scope of the social construction, and therefore the degrees of ontological and epistemological constructivism. Are discourses and social constructions constitutive of our reality (ontological constructivism) or is there a reality in itself *(an sich)* that we are unable to experience – as we will always understand and experience the world through predominant ideologies or cultures – but which exists nevertheless (epistemological constructivism)?
- The second axis is an extension of this and is about scepticism and relativism. As discussed in Chapter 7, we can talk about both local relativism and universal (global) relativism. Local relativism sees the truth about the world as rooted in historically and culturally defined rationales. Universal relativism is reminiscent of scepticism. It argues that, since all we can say about the world is coloured by our historical and cultural place in it, there is no external, non-linguistic source of verification for our pronouncements about the world. As such, there is no way to judge whether one thing is more correct than any other.
- The third axis concerns the choice of analytical focus. The axis can be drawn between micro-oriented analysis strategies that are aimed at practical everyday knowledge, and macro-oriented analyses of institutions and general discursive formations. Where the subject field of a micro-oriented analysis strategy automatically leads to greater focus on how actors participate in the creation of social constructions, macro-oriented analysis strategies tend to see the actors as determined by discourses and constructions.

The questions we must therefore pose about social constructivist analyses are: What is it that is constructed? Is it the actual phenomenon? Is it the idea of the phenomenon? Is it the way the phenomenon behaves or how it looks in a specific context?

Is it the experiences that result from the phenomenon? These questions are raised, in a very sensible and thoughtful way, in Ian Hacking's book *The Social Construction of What?* (2000). In it, he discusses what is special about social constructivism, and in particular, points to its – at the outset – liberating effect and its ability to question conditions that otherwise appeared immutable. Social constructivism is about how power is established, exercised and transformed through language and signs. This means that power is basically immaterial. This is not the kind of power exercised by the police when they clamp down on a demonstration or arrest criminals, but power in a more abstract sense, as the boundary between what is normal/abnormal, inside/outside or legal/illegal.

Hacking identifies three basic assumptions, which in his opinion form the basis for social constructivism's way of thinking (Hacking 2000: 6). X is the phenomenon being studied (e.g. gender, race, nation, body, sexuality, family):

1. X does not need to exist or have the form it has. The way X looks to the world is not determined by its nature and is therefore *not* immutable.

Often, assumption 1 is followed by assumptions 2 and 3, which imply criticism of no. 1:

2. X is bad.
3. We could be better off if X were not there, or if X were radically transformed.

According to Hacking (2000: 15), social constructivism is also based on the underlying premise that the phenomena studied are generally regarded as immutable by others in the context in which they occur – the social constructivist researcher aims to show that the phenomena others consider truisms are not actually truisms. There is, therefore, a point that precedes the three above:

0. In the current situation, X is taken for granted as a natural fact.

This point is often used to emphasise the special nature of social constructivism, but the problem is that it is not always a faithful representation of other perspectives. In reality, hermeneuticists, phenomenologists, positivists and pragmatists are not exactly blind to the fact that certain rules and norms are created socially. One criticism of social constructivism, therefore, is that it attacks others for positions that they do not actually hold.

The social in social constructivism

Peter L. Berger and Thomas Luckmann's book *The Social Construction of Reality* (1966) introduced the concept of social construction and is widely considered as the precursor to the social constructivism of today in scientific circles. They were heavily influenced by phenomenology, especially Alfred Schutz's sociological phenomenology, which revolved around day-to-day life, but also by both Karl Mannheim and Marx's sociologies of science. Berger and Luckmann can, therefore, be seen as forerunners of the more micro-oriented aspect of social constructivism. In the book, they stress that even basic and absolutely ordinary everyday knowledge is created and maintained on different levels of social interaction. Their focus on everyday experiences and everyday knowledge expanded the sociology of knowledge, which progressed from exclusively dealing with ideologies and scientific disciplines to being a key field in sociology, and was formulated as a broad sociological theory of society that observed and analysed knowledge based on what specific communities considered to be knowledge. Berger and Luckmann built upon five fundamental insights (Berger & Luckmann 1966):

1. Knowledge always springs from a specific position or social place.

2. Human consciousness is determined by its social being and its place.
3. What is true for some may be false for others.
4. Social facts or institutions must be considered and analysed as things, rather than as explanatory factors.
5. The sociology of knowledge must work with everything that is considered knowledge in a given society.

Berger and Luckmann are interested in uncovering how all kinds of knowledge are created, and how individuals and their awareness of the world are constructed in repeated interaction with other individuals. They highlight the fact that our understanding of reality stems from our presence in the social world, and that the social world plays a key role in the definition of this reality (Berger & Luckmann 1966). Reality is, therefore, always created through interpretations of experiences that are not individual but negotiated and fixed in social and local contexts.

According to Berger and Luckmann, the categorisations and typologies we use to describe and define our social reality are not only forms of language, but also depend on our practical, everyday life and surrounding institutions. This means that society and its institutions, which for the individual constitute an objective reality, are human products of recurrent patterns of behaviour and meanings that are created through everyday interaction and typologisations, (re-)established in everyday habits, routines and general understanding of our own and others' behavioural patterns. Every individual is born into a society that has emerged from the struggle between different meanings and perceptions of reality in historical and socially specific contexts. For example, at different times, the importance of women and men, or of poverty and wealth, may appear different.

According to Berger and Luckmann, struggles are played out in a manner similar to a pendulum. First, there is an *externalisation*, by which the construction and reproduction of society and its institutions through habits, routines and interpretations become everyday practice. This is fixed over time, and therefore the constant negotiation and struggle with regard to habits, routines and interpretations is *objectified* into objective and unchangeable facts, norms and institutions. The norms and typologies are subsequently *internalised* by the individuals who make up the society and use the institutions.

For instance, the typologisation of what a woman was in 1850s Denmark, and the role she could/should play in society, was created through historical and culturally defined social relations and power dynamics in that society. It was created through habits, routines and interpretations that had been externalised through society's various institutions. As a result, the franchise was not extended to women by the June Constitution in 1849, which otherwise enfranchised citizens who did not receive financial support from the state or local authority. In this case, the knowledge that placed women outside the category of "citizen" was constructed by the power relations that had been established in the everyday habits, routines and interpretations, and through, for example, medical, literary and philosophical categorisations. In this light, the typology of women was objectified and subsequently internalised in society's institutions – and most women did not question the fairness or the supposedly natural state of their lack of voting rights.

According to Berger and Luckmann, this process is not passive, it involves the ability to change categorisations and meanings. This happened, for example, when women in Denmark were granted the vote in 1915, after approximately 25 years of fierce struggle for the right to self-definition (and hence new habits, routines and interpretations).

For Berger and Luckmann, the construction of social reality affects the subjective experience of social reality through the process of internalisation. Even though, in principle, they do not consider institutions to be determinant for the individual, they describe (as do many other social-constructivist thinkers) the internalisation of institutions as a reasonably stable and solid order, which does not leave much room for change through the individual's actions.

Knowledge and language in social constructivism

A more radical interpretation of social constructivism is espoused by the American psychology professor Kenneth Gergen (1934–), who emphasises that the subjective experience of thinking and knowledge is never individual but always constructed and embedded in social relations, and can only be observed collectively. According to Gergen (1994), we only interpret the world and create knowledge through collective discourses and institutions created by society. Experiences are communicated and understood exclusively via language and its categorisations. As such, Gergen refutes the idea of the individual as an autonomous whole, and stresses the constructive role of language in the creation of knowledge. There are no individuals or individual experiences independently of how the collective defines the individual. The individual's identity has no essence and therefore cannot have individual experiences – rather, it is created only by the collective institutions in its relationships.

Therefore, Gergen is interested in studying *multiple identities*, the establishment of which depends on the social, cultural and historical space in which we find ourselves. In Gergen's understanding, the ego, the psyche and the self are replaced by language, narratives, culture and history. The individual or self does not create language, narratives, etc. Instead, narratives, culture and history create the individual and the individual's subjective experience of social reality. Where Berger and Luckmann's focus is on the more tangible and everyday social construction of reality, Gergen starts with a bigger picture, one in which knowledge and experiences of, and about, the world are always created through a pre-established language filter, which only changes as a result of struggles about categories between social groups with different interests.

Discourse analysis

The discussion of the relationship between language and the world, and the perception of language and meaning as constructing the social world, are pivotal to discourse analysis. Discourse analysis takes as it's starting point that language and speech are never neutral representations of the world. It is not the world that asks to be represented in a certain way, but the representation of it that makes the world specific. For example, there are no natural features (essences) in Muslims or homosexuals that lead to them being represented, described and discussed in a certain way – rather, the groups concerned are assigned a specific place and specific options for action in society by virtue of the way in which they are represented and understood. This realisation can be difficult to fully understand, since in everyday practice it seems as though there is a connection between phenomena and the way in which they are represented. This is because legitimacy and credibility are also established through discourse. Phenomena therefore attain a certain social position through discourse, which also means that this position will seem natural and true. Discourse analysis does not contain any truth outside of the discourse – even the truth effects emanate from within the discourse.

The different interpretations and gradients of discourse analysis can be placed along the same three axes as described for social constructivism: ontological/epistemological constructivism, local/global relativism and micro-/macro-analytical approaches.

Discourse analysis can be described as post-structuralist, in the sense of an extension of the structuralism of, for example, Saussure (see Chapter 6). Just as in Saussure's sign theory, post-structuralism is based on the fact that the meaning of the sign does not come from an inherent essence, but is established through the relation of differences and similarities it enters into with other signs – the sign itself is arbitrary, as there is no natural relationship between expression and content. In Saussure's sign theory, this results in the system of meaning being considered as an immutable and centred structure that gives individual structures a fixed place and determines the elements unequivocally. Post-structuralism, however, stresses that all structures are decentred and fluid. Metaphorically, Saussure's theory of signs is like a fishing net, in which the individual knots all have their specific place (essence) by virtue of their position relative to the other knots. In a post-structuralist perspective, the knots are more akin to postings on the Internet. Internet postings are not defined and set in stone within a specific corpus, but move constantly due to other posts that refer to them being published or deleted. In this sense, meaning is a fluid motion – it is not fixed, but arises and is confirmed in the concrete situations in which it is articulated.

In discourse analysis, language is considered to be an event out of which a specific meaning emerges. Of a discourse, we can say that it:

- is time-bound (to the moment at which it is used)
- stems from a user (it is not real until it is used)
- must credibly be identifiable with the world to which it refers (tied to specific situations)
- is always directed at someone (only by communication and practice does it become real)

Unlike in hermeneutics, there is no underlying, hidden meaning to be construed from between the lines of the statements being studied. In discourse analysis, meaning is established on the text's own level and through the way that words, concepts and themes are related to each other, both in the specific text and in the general text universe of which the specific text forms part. The father of deconstructivism, Jacques Derrida, describes this in his famous statement "il n'y a pas de hors-text" or "there is nothing outside the text" (Derrida 1976: 158f).

One common mistake associated with understanding discourses consists of the *intentional erroneous conclusion*. Here, a discourse is reduced to an expression of the speaker's intention and is nothing more than a function of subjective intentions. The other common mistake is referred to as *realistic conclusion*, by which the subject is reduced to the mouthpiece of a discourse. This serves as a barrier to the insight that discourse both does something to the object and sets legitimate subject positions (i.e. legitimate positions from which to talk). Unfortunately, these mistakes often occur in studies that call themselves discourse analyses. It is crucial that discourse analysis manages to reflect on language, subjects and objects, at one and the same time. Someone talks to someone about something, but the speaker is not in full control of the discourse – the discourse itself restricts the opportunities for making a statement, and acts as a regulatory mechanism in relation to establishing both the object and the subject positions.

Language, knowledge and power

The connection between language, knowledge and power is fundamental to the thinking of the French philosopher and historian of ideas Michel Foucault (1926–1984). As the *grand old man* of discourse analysis, Foucault has had an overwhelming influence on the humanities and social sciences over the last several decades. He was interested in the effects of power and knowledge and how they discipline and regulate subjects. He conducted extensive studies of how historical systems of meaning have institutionalised the relationship between knowledge and power in particular epochs of cultural history, and which are reflected as dominant social practices that we take for granted. Foucault studied such varied phenomena as prison design (1975), hospitals, madness and civilisation (1961), and health, illness and sexuality (1976–1984).

Contrary to the Marxist concept of power, defined as a material relationship between the oppressor and the oppressed, Foucault does not consider power as something that actors can either be or not be in possession of, or as something negative and destructive. He sees power as a creative force that exists as a basic structure in all social relationships. Power normalises and disciplines individuals through different administrative techniques that are used to exercise control and normalise. He describes this in his historical analysis of the changing definitions/discourses that constitutes what is a crime. According to Foucault, a specific discourse leads to specific forms of punishment, reflected in the prison's physical design and the treatment of offenders. According to the dominant discourse, prisoners are locked away, have the evil driven out of them, or are re-educated or civilised through self-management (Foucault 1995).

For Foucault, power is productive because it shapes social relations – and it is in this sense that he can be described as a social constructivist. It is power's relationship with knowledge that gives rise to its productive effect. Knowledge is not, and cannot be, neutral – all knowledge processes describes the world in certain ways. The description itself forms the object being described; whether it is about "the Middle East" or "depression", the form of knowledge is a way of describing the object, which then defines the part of the world being described. The description draws boundaries for what is inside the object and, of course, what is outside.

A Foucault-inspired discourse analysis examines how objects are established, what rules are applied and, possibly, the causality embedded in the definition of object and the hierarchy this establishes. "The Middle East" is defined both as a geographical, political and religious place with some specific characteristics (e.g. geographic location, lack of democratic regimes, Islam as the dominant religion). Thus, the concept includes or excludes – depending on which characteristics are used as the main delineators – countries such as Israel, Turkey and Algeria, each of which has several different parameters. Likewise, with "depression", which is assigned certain characteristics (sadness, apathy, lack of zest for life, trouble sleeping, perhaps suicidal thoughts), as well as causes of a physiological, psychological or social nature. Both the causes and the characteristics create discursive formations that regulate the subjects found within and outside the boundary that the definition draws around the object.

Foucault's concept of discourse aims to analyse how power

and knowledge are coupled together and to regulate practical actions on an institutionalised, collective basis. Discourse is the regulatory mechanism and, therefore, in different contexts, an order that expresses an epoch's shared knowledge. This endows the words with collectively binding validity. It becomes the obvious, but not necessarily conscious, basis for the actions and for what activities are possible and permitted in the specific epoch. The power of discourse consists, therefore, in defining and excluding the incomprehensible or unacceptable, and allocating roles to the discourse practitioners within hierarchies in order to maintain those same hierarchies.

It is important to emphasise that the term "discourse" does not only refer to a way of speaking. It should be understood more generally as a delineated body of statements that are formulated and may contain all sorts of texts in the broadest sense. No kind of text is more privileged than others. The same archive may include patient records, psychiatry textbooks, rulings by the patients' complaints board, slang, derogatory jokes, Hollywood movies and state policies about psychiatry. In this context, the discourse forms the regulatory mechanism that helps to define, for example, depression as a primarily physiological and neurological problem (object) with associated treatment methods, as well as the subject-positions that entails (i.e. who can talk of depression and when, and how you can be, or not be, depressive). The formation of the knowledge object results in a definition and specification in relation to other discourses, such that, for example, depression is understood through its contrast to psychological disorders and its similarities with other, more physiological conditions. The use of concepts, words and themes (e.g. neurology) linked to the discourse (in the case about depression) helps to establish power relations concerning who has the right to speak about the object and in what situations. This means that not everyone is able to speak with authority and legitimacy about depression, and not all subject positions can be adopted in all situations – they are defined via the discourse. The statement position is defined through the discourse and establishes hierarchies – i.e. neurologists, psychiatrists and doctors are higher up the hierarchy than psychologists or patients, so their statements are accorded greater significance.

Foucault points out that several different discursive formations are often present at the same time, and that these are embedded and partially disconnected in contradictory layers in the given epoch. He therefore deals mainly with the exclusion or inclusion of phenomena on the periphery of a society and a period's norm systems (e.g. the insane, criminals, the sick) in order to better identify the regulatory mechanisms that are at play when defining objects and the legitimacy of subject positions.

He uses two different analytical strategies that are bound together in both space and time: the archaeological and the genealogical.

- The *archaeological* analysis is conducted through a study of the rules for meaningful and true statements in an epoch. A *synchronous* analysis within the same epoch explores the regularity and the spread of the discourses – in other words, how often they occur and the extent of their range. For example, the neurological discourse looks at how often certain forms of explanation and idioms are used and to how many different areas they have spread. Is it only in the psychiatric system that the discourse about depression occurs, or is it also found in other areas, e.g. the psychological, educational or legal field? The archaeological survey reveals which knowledge regimes – i.e. historical rules – determine what is regarded as true and legitimate in an epoch.
- The *genealogical* analysis is *diachronic*, as its purpose is to trace the continuity and discontinuity in discursive formations. Unlike historical and hermeneutical approaches, the purpose of the genealogical analysis is not to study past events/phenomena in order to identify causal links, path dependencies or origins that serve to explain how a phenomenon has become what it is. Foucault's genealogical

analysis goes the other way – it examines how current regulatory mechanisms resemble or differ from past regulatory mechanisms, with reference to themes, concepts, words and subject positions. The aim is to identify the struggles that have taken place and that have affected which discursive formations have been retained and which have been marginalised. Traditional historical analysis studies the origin of phenomena. Such an analysis will marginalise everything that does not resemble or link to the original. Genealogical analysis, on the other hand, challenges the idea of origin. It looks at how the practices we take for granted today (and hence the critical perspectives) have become institutionalised through discursive power struggles. The genealogical approach, therefore, does not marginalise practices and struggles from a perspective of origins, but is based on contemporary practices, and delineates the analysis on that basis.

Sociology-inspired discourse analysis

Another widely adopted approach to discourse analysis is rooted in a more sociological-linguistic perspective and is formulated by two post-structuralist political scientists, the Argentinian Ernesto Laclau (1935–) and the Belgian Chantal Mouffe (1943–), who formulated their perspective in the book *Hegemony and Socialist Strategy* (1985), based on a fusion of Marxism, structuralism and post-structuralism. In it, they develop Saussure's concept of signs by placing greater emphasis on the fundamental instability of discourse, which leads to ongoing discursive changes. Laclau and Mouffe point out that all discourses are established in the fields of discursivity (1985: 111), which are characterised by an infinite diversity of meanings that the discourses can adopt but never fully cover. For example, the phenomenon of "woman" has an enormously broad and mutually contradictory cluster of meanings, which no discourse will ever be able to accommodate. The discourse is therefore unstable, and there will always be controversy about how exactly it should be formed. A discourse therefore consists of semi-fixing the discursivity. Accordingly, there is always something else outside of any specific discourse. This makes the discourse unfinished, and therefore something that is changeable and moveable. This instability opens up the question of how certain elements in the discursive field are linked relationally with each other and become a discourse.

This might sound as if there is a fundamental difference between Laclau and Mouffe's discourse analysis and the idea that there is nothing outside the text, which I touched on before – in one version, there is always something outside of the tangible discourse; in the other, there is nothing outside. However, it is important to understand the precise meaning, which is the same in both cases: in Laclau and Mouffe's interpretation, there is always something outside the discourse that causes instability, but whatever is outside is not essential and does not determine what the discourse will consist of. The discourse is a constant struggle about meaning and about the specific shape of the discourse. The specific meaning arises out of the fixing of the discourse – not because of anything outside it.

Laclau and Mouffe use the term "hegemony" for the dominance of a particular discourse over a limited period. The discursive struggles and various groups' struggle for hegemony are central to Laclau's and Mouffe's discourse-analytical interest, which was basically all about defining a new socialist project in the mid-/late-1980s, when socialism was in crisis.

The discursive field contains, in Laclau and Mouffe's terminology, *empty signifiers*. These are concepts that are not clearly and tightly defined, and the struggle for hegemony is largely about winning the right to determine their meaning. The precise meaning of concepts like "growth", "welfare" or "democracy" is of the utmost importance for which politic is considered to be the right one. In order to win the discursive struggle for these concepts (the empty signifiers), political groups must determine their meaning, via articulation. The various groups attempt to define the empty signifier because – for example, depending on how its meaning is defined – "growth" can give rise to tax cuts, tax increases, more redistribution, less redistribution, etc.

The definition of empty signifiers can establish what Laclau and Mouffe call a *nodal point* in the discourse. A nodal point is a central concept or a central contradiction that holds the discourse together. Around this point, a number of other words and terms coalesce and make the nodal point's attribution of meaning more precise. In the case of "growth", such terms would be *accountability, public spending* and *tax cuts*. Nodal points are always fixed in a temporary and incomplete manner. Therefore, discursive struggles always try to strip the nodal points of their meaning and (re)fix the empty signifiers with new content.

Establishing subject areas in social constructivism and discourse analysis

The starting point for discourse analysis and social constructivism is that our knowledge about what we perceive to be truisms and natural (e.g. the body, the nation, society and the individual) is not based on the phenomena's inherent essence, but that we consider them representations of the world that are established and embedded in specific cultural and historical contexts. This means that they are *contingent* or random and therefore need not be as they are. Discourse analysis and social constructivism position themselves as anti-essentialist, and focus more on the process by which the constructions/discourses emerge – and on the marginalisation of other meanings and positions that takes place in this context – more than on what the fixed structures look like. Unlike the structuralist perspective, the social constructivist and discourse-analytical perspectives consider the meanings of phenomena to be constantly moving and fluid.

Since we only have access to the world via representations of it, language becomes central to the establishment of the reality we share. This makes interpretations, meanings and observations absolutely dominant as explanatory models – and also, therefore, as a field of study. Knowledge of the world is created and maintained in social processes that entail a continuous struggle about definitions. For this reason, there may be definitional, discursive struggles over how the world should be seen and phenomena understood. The way in which we agree to define and identify the world is central, because the way we understand the world affects our actions in it. Taking illness as an example, the definition and identification of illness affects how people behave as health-care professionals, as patients and as relatives, how people experience being sick and how social institutions deal with the phenomenon (e.g. sick pay, hospital design, nursing auxiliaries and health insurance).

The social constructivist and discourse-analytical perspective is basically critical, and seeks to illustrate the production of truisms and the constraints that they place on the way in which we understand and organise ourselves and each other. The starting point for the critical perspective is that, since truisms are constructions, they can be identified, and therefore replaced by other – possibly better – structures. Even though the critical perspective deconstructs and reveals strategies for manipulation – for example, how identity is linked to different consumer goods in order to induce a greater desire to spend without most consumers being aware of it – it can also be used the other way around. Thus, it is possible to work on how, strategically, to link constructed truisms such as race, nation and gender with your own needs, and in doing so generate extra sales of certain products or cause subjects (e.g. students, teachers, the sick, the old, teenage mothers) to behave in certain ways.

Social constructivism and discourse analysis are not interested in the individual's intentions and experiences, because the individual does not represent him- or herself and does not have an individual essence. The individual's emotions, actions and experiences are considered to be representations of general discourses, social constructions and structures over which they do not exercise control. As such, there is not much individual agency in social constructivism and discourse analysis, which take after structuralism in that sense. Several social constructivist-inspired writers (e.g. Pierre Bourdieu and Anthony Giddens) have tried to incorporate thinking about agency into social constructivism. However, their analyses have centred not on the individual, but on individuals' collective uses of established positions and categories.

Examples of research questions

Based on the social constructivist and discourse-analytical perspective, let us pose the following research questions about the World Bank's anti-poverty work:

- How does the World Bank construct anti-poverty work through its aid practices?
- How is poverty established as an object in the World Bank? To what legitimate recipient and donor positions does this lead?

The starting point for both approaches is that poverty does not have an essence per se. Poverty is something we define and establish arbitrarily, depending on the interests/discourses we take as our starting point. As such, discourse about poverty is not merely a representation of how something is; the representation has an influence on the world. The social-constructivist approach allows for both a macro- and a micro-oriented analysis. The macro-oriented analysis could be directed towards the institutional frameworks and general structures of poverty and anti-poverty action. A micro-oriented analysis will focus more on how knowledge – and thus constructions – are established in everyday contexts through the routines and traditions associated with the categorisation of countries as poor and worthy of receiving aid from the World Bank. Such a study could examine how poverty is defined by different groups in the World Bank, how different countries are typologised and, by extension, the explanatory models for poverty. The object of such a study could be the struggles surrounding interpretation within the World Bank's anti-poverty work in the period 1970–2000. In this context, you could try to identify which professional groups within the bank (e.g. environmentalists, sociologists, lawyers or economists) have played a role; what their interests have been in this connection (e.g. to strengthen their own position in the World Bank, to create a need for further recruitment from the same profession, professional logic, etc.); and what the different definitions and categorisations mean for anti-poverty work. This could involve a study of the consequences of the ongoing renegotiation of legitimate categorisations of poverty and, by extension, the causes of and therefore solutions to poverty problems.

The discourse-analytical research question could also be explored through both a micro- and macro-oriented analysis. At the macro level, a discourse analysis could focus on how poverty is established discursively through themes, concepts and words. Since discourses always consist of someone saying something to someone – and thus seek to regulate subjects – discourse analysis could, in this context, uncover what, in a specific situation, the discourse means for the legitimate position and options for action of the recipient and donor countries, as well as critics and supporters. Since discourses are continuously negotiated and fixed via relations of power and resistance, the conceptual struggles will be particularly central in this perspective.

Knowledge creation in social constructivism and discourse analysis

Social constructivism and discourse analysis are interested in understanding how specific meanings/structures and discourses emerge and how they shape our ways of acting in and representing the world. Since constructions and discourses always take place in historically and culturally specific contexts, they cannot be studied in isolation from them. We therefore describe the processes that take place in detail. Typically, the analysis will focus on the points at which the constructions and discourses are seen to move and become the objects of struggle. For this reason, Foucault deals with phenomena that are outside of or peripheral to the norm (the mentally ill, sexuality, criminals). This provides clearer access to identifying and shedding light on that which is considered to fall within the frameworks of normality. The normal is seldom formulated, precisely because it is normal. To study these processes, it is important to have an analytical eye that focuses on the unstable and fluid. Although social constructivism and discourse analysis aim to study how constructions and discourses emerge and are used, they are also controlled by the underlying assumption that knowledge is always created in the social sphere and is therefore explicable by describing how this social process takes place.

Discourse analysis and social constructivism use both inductive and deductive approaches, preferably in mutual extension of each other. Processes are based on the theoretical assumption that we have access only to representations of the world, which in turn has consequences for how we see it and act in it. The study of processes, and the understanding of knowledge as something contextually embedded, requires that the analytical work is based on the specific situation and process in which the discourses are established and negotiated. As such, the analyses are ideographic based on very specific texts, events and contexts, but the abstract outcome of this can, in principle, be generalised to other contexts.

Knowledge is understood both contextually and as a regulatory power, i.e. there is a difference between what people in different contexts and times consider true, depending on the constructions and discourses that dominate their social space. This means that the researcher's knowledge must be subjected to the same scrutiny as the context being studied – because the researcher's context also affects the types of knowledge that are possible. This involves varying degrees of (local/global) relativism in relation to the knowledge that the researcher is able to build up. The detailed description can be seen as a way in which to establish credibility in the analysis and create coherence between what is being described and the conclusions reached. Therefore, a conclusion is credible if it fits in with the system of the other statements on the basis of which you reached the conclusion in a manner free from contradictions – in other words, a coherence-based concept of truth.

Example of data acquisition and processing

In our social constructivist example on the construction of anti-poverty work, it would be obvious to look at how (everyday) knowledge, routines and practices in the World Bank are based on the different professions' typologisation of poverty and their struggles for the right to define it. The purpose of the study would be to understand and explain the continuous process to which the struggle over meaning gives rise. You could collect material about the different routines and descriptions/explanations of anti-poverty work over a number of years, or base your study on a particular conflict and examine the struggles over meaning that took place in this context – for example, the famine in the South Sahara in the 1990s. In this context, you could look at how the different groups described, talked about and understood the disaster, how they perhaps used the meaning assigned by NGOs or donor countries, and which routines were subsequently institutionalised. The study would typically be inductive (apart from the underlying theoretical premise that the social context defines what is true). Although the starting point is the construction of poverty that is dominant and institutionalised in a specific context, you would need to examine specifically (inductively) which struggles about which principles and routines took place in order to explain how the constructions emerged, and what effect they had on institutions and meanings.

In the discursive analysis of poverty and legitimate recipient and donor positions, you could examine how the border between what is outside and what is inside the discourse about poverty is established. In this respect, it would be obvious to refer to both internal and external documents and policies on the aid principles, and political discussions and definitions of poverty. You could examine the specific concepts and words used in the description of poverty, as well as the themes that are linked to poverty reduction and the definition of legitimate recipient countries.

There are various strategies for conducting discourse analysis – some of which are more functionalist than others. Whatever approach is adopted, the key is to look at the knowledge creation that occurs when meanings are fixed. In our case, this may involve both interpretative analyses and numerical quantification of the spread. It would be relevant to analyse the World Bank's official reports for the period 1970–2000, the World Bank's internal policies, the negotiations with donor and recipient countries, etc., in order to identify similarities and differences in the way in which the language used produces the object and corresponding subject positions.

CHAPTER 9

Pragmatism

Nowadays, pragmatism represents a return of the individual and an emphasis on practice. In contrast to, for example, structuralism's focus on abstract structures or positivism's focus on revealing generalities via objective observation, pragmatism aims to explore how specific individuals act in tangible situations. The attention is on why specific, tangible situations appear the way they do and not so much on whether certain conditions apply more generally.

Pragmatism has its origins in the work of the American philosophers Charles Sanders Peirce, William James and George Herbert Mead, all of whom wrote their main works around the start of the 20th century. Practice and action are key elements in their understanding of science. This has had significant implications for thinking about the individual, society and communities.

Peirce was the first to formulate pragmatism's maxims, in 1878. They form the basis of much later pragmatic thinking, and were most recently reinterpreted by Richard Rorty (2007). Pragmatism sets evaluation criteria for scientific statements that are radically different from the other theoretical perspectives we have looked at in this book. Pragmatism is not a new approach but, since the 1970s – and especially in the last two decades – it has become increasingly important. Today, it plays an important role in cultural disciplines, semiotics and pedagogy, as well as organisational research. In the latter in particular, interest in, and work with, real-world practice has grown considerably in recent decades, and pragmatism and its principles have been highlighted as a useful approach.

Like other theoretical perspectives, pragmatism has developed over time and, as such, encompasses several threads that branch out in many directions. It cannot therefore be defined as an actual school. However, it does have certain general characteristics:

1. Pragmatism is basically a challenge to the many dualisms in Western science, particularly René Descartes' famous distinction between body and soul, which forms the basis of much post-Enlightenment philosophical thinking, i.e. from the 1700s onwards. Rather than placing knowledge in the soul and rejecting the body and its senses as potential sources of error, as Descartes did, pragmatism insists that all knowledge arises out of physical sensations in specific situations.

2. At pragmatism's core are human action and experience. People are considered active participants in the social world, which they affect and form through their practices. The knowledge-constitutive interest is usually *ideographic* and often originates in the micro level. Pragmatism focuses on how experiences of previous situations affect, and are applied to, current actions, and the potential consequences of this.

3. Pragmatism cannot be defined as exclusively realistic or constructivist. The individuals involved in social phenomena also interpret them, but they do not have total freedom to do so. The phenomenon occurs in a real situation, which places limits on the possibilities of the interpretation. The focus is on the importance and consequences of social actors' actions. As such, the meaning of a phenomenon is fixed by its consequences. This entails adopting a processual view of the phenomena studied – what is interesting is not so much what they are, but how they emerged with a particular meaning. Pragmatism's challenge to Descartes' distinction between body and soul means that it cannot

be defined as either idealistic or materialistic. It considers body and mind to be the inseparable common basis for human actions.

4. Epistemologically, pragmatism assumes that all knowledge has a bodily sensation as its starting point and is achieved through the interpretation of signs that represent the world. This analytical practice is based on *abduction*, which combines aspects of deduction and induction. Abduction is also known as "a qualified guess". Using analogue links between previous knowledge and experience and the phenomenon we seek to understand, we create opportunities to understand and identify unknown objects. The concept of abduction is central to the pragmatic analysis and is described in greater depth below.

5. Since all scientific knowledge is based on the researcher's interpretation and qualified guesses, the question of whether knowledge can be *value-free* is irrelevant in pragmatism.

6. Its view of human nature is one of pragmatism's most distinctive characteristics. It is developed as a response to utilitarianism, in which the understanding of human action is teleological. In other words, the way in which people act is directed towards a specific goal, determined in advance, i.e. to achieve the greatest possible benefit for the individual and others. Unlike utilitarianism, pragmatism focuses on the consequences of human actions, because only in this way is it possible to understand their meaning. Actions cannot be directed at a specific purpose, or be determined in advance, since the actual situation, and the past experiences applied to the situation, are bound to influence the action. Pragmatism considers human intentions to be processual, relational and situational – and therefore both individual and social at the same time.

7. The pragmatic truth theory is based on a *relationship* between people, institutions and statements. This means that we consider something to be true at the end of our inquiry, and that our results are helpful and useful ways of explain-

ing phenomena and events. In other words, it is not a requirement that our knowledge must be true in any objective sense, as long as it serves to help us understand the truth of the concrete situation studied.

The roots of pragmatism

The characteristics of the study

Charles Sanders Peirce (1839–1914) has been called the father of pragmatism. He studied logic and mathematics and had a degree in chemistry. He wrote and philosophised on many different subjects, from logic, language and communication, to learning, mathematics and his highly developed theory of signs, of which almost nothing was published during his lifetime. His influence stems mainly from his students and colleagues, who built on and referred to his ideas. Their publication of his collected papers accounted for his subsequent influence. Peirce is particularly well known for his development of pragmatics and for his theory of signs (semiotics). In contrast to Saussure's theory, Peirce's semiotics is not based on language signs, but encompasses all kinds of sign systems, in order to say something substantial and practical about the relationship between the knower and the known.

As mentioned, not much of Peirce's work was published during his lifetime. However, two articles from the magazine *Popular Science Monthly*, "The fixation of belief" (1877) and "How to make our ideas clear" (1878), reflect his interest in, and concern for, thinking about truth, scientific study, and how we can come to know the unknown.

Peirce points out that new knowledge usually emerges when our existing knowledge and habits prove inadequate to explain and understand a new experience. In order to gain new knowledge, it is important that we are what Peirce calls *fallibilists*. We must, in other words, possess the ability to accept anomalies that challenge our current beliefs. In Peirce's terminology, beliefs serve as the basic framework of understanding that makes the world intelligible to us. This does not mean that belief is necessarily true, but it is useful – it allows us to act and understand the world around us *as if* it were true. Doubt and wonder arise only when our faith is challenged to such an extent that it no longer helps us create understanding in specific situations. Peirce thinks that people generally find it difficult to live with doubt, uncertainty and the inability to understand what is going on, and therefore try to eliminate doubt by creating a new faith with which to grasp the unknown and incomprehensible (Peirce 1877; Cunningham, Schreiber & Moss 2005). He identifies three ways to do this (Peirce 1877):

1. The *authoritative* method of reaching conclusions vanquishes doubt by acquiring new authoritative explanatory frameworks from credible authorities (e.g. if an inexplicable natural phenomenon such as a tsunami is presented by a priest as God's wrath, and this is accepted because he is a religious authority).

2. The *a priori* method of reaching decisions is based on current practice, of which we all avail ourselves in situations that are not instantly understandable. Based on our general knowledge, we establish links between the immediately incomprehensible and our current understanding. This, of course, helps extend our faith, but does not enable us to understand anything that is radically different from our current framework of understanding. The difference between the *authoritative* and *a priori* methods of reaching conclusions is that, with the latter, we are not presented with a new faith that we accept because it comes from authority, but we expand our existing beliefs to accommodate the otherwise inexplicable experience (e.g. we understand the tsunami as an underwater storm).

3. The *experimental or abductive* method of reaching conclusions, which is not used until we find ourselves in situations where the incomprehensible experience can neither be made understandable nor be accommodated within our

general knowledge/belief system. Here, we need to develop new basic structures for understanding with which to process the new experience. Peirce calls these situations experiments, which he thinks are controlled by abduction. In this context, a tsunami can be understood by reference to its elements and the way in which they share ideas, patterns, conventions, functions or effects with other elements we know from multiple different contexts. An abduction always has to be tested for its durability, through a study of how qualified and relevant the guess is.

Abduction

The concept of abduction is the key to pragmatism. A study based on abduction observes and uses all signs, especially small and not immediately significant clues from the incomprehensible situation's context, and requires a detective's flair for establishing hypotheses. Abduction heralds a new way of working, in which it is possible to identify an unknown object whose being cannot be proved, but is possible. The interesting thing about abduction is not, therefore, that it proves that something is true, but that it seeks, in a creative way, to say something about the world that reveals new or unknown phenomena. Whether what we say is true or not must then be considered more systematically by testing the possible hypotheses against relevant material. We also know abduction as Sherlock Holmes' preferred way of working.

Peirce's idea of abduction is an extension of both deduction and induction. Deduction is the movement from theory to result. It therefore serves as a theoretical framework for arranging observations, assuming that the method by which we reach conclusions is compelling and necessary. In other words, we study whether the theoretical hypotheses actually hold when we look at reality. In contrast, induction is a movement from result to theory, a movement that remains incredibly close to the source and has difficulty rising above the source of the knowledge. In other words, we look at reality and use our previous experience to develop a probable theory.

Abduction therefore combines elements of deduction and induction to produce one or more largely plausible assumptions about correlations, or what Peirce himself calls "qualified guesses" (Peirce 1932). Abduction is similar to induction because it uses the encounter with reality in order to develop new theories about the world. However, we do not just follow a linear logical progression from part to whole (theory). Rather, we establish many hypotheses to which the encounter with the world gives rise, based on previous experiences of the world. All these possible answers and theories are then tested again on the material. This stage resembles deduction, because it involves testing of the probability of the hypotheses.

This is also inherent in the designation "a qualified guess", where "qualified" refers to the prior knowledge and experiences that allows us to establish multiple useful hypotheses about the world, while "guess" shows that we are seeking to gain new knowledge that does not follow a strict logic. The crucial difference is that abduction implies a less formal and more immediate approach to the world that focuses on obtaining new knowledge of specific situations and phenomena. Abduction, induction and deduction all play important roles in Peirce's universe, but abduction is the only method of reaching conclusions that allows us to identify a truly unknown phenomenon in the light of the knowledge we already possess. Truth is not universal, therefore, but is related to – and therefore changes with – specific situations. Abduction is the first step toward truth.

The concept of truth in pragmatism

The thinking outlined above about studies and abduction also has consequences for the pragmatic understanding of truth. To this William James (1842–1910) plays a particularly key role. Peirce was not generally acknowledged at his time, nor did he have a tenured post at a university, so initially it seemed that James was the main pragmatist thinker, despite the fact that he constantly pointed out that his work was based on Peirce. James discussed the concept of truth he observed in the research of the day, and came to the conclusion that the use of truth was limiting because it was closely associated with the fear of making mistakes and with attempts to maintain established knowledge. He believed that truth should serve a purpose (James 1907: 222) and, like Peirce and Dewey, he associated it with the concept of experience. James considered ideas true if they were able to understand and explain people's ongoing experiences (James 1907: 34).

The consequence of the pragmatic theory of truth is that the results at which we arrive should not be regarded as final and immutable. Our results must be the most useful, probable and credible we can arrive at, based on the situation and the information we have at the given time and the context in which we formulate our results. Deciding how useful they are may entail deliberations about whether theories and conclusions describe and explain the phenomenon we are studying better than if we had not applied them. To evaluate this, it is vital that we are able to relate to them as fallibilists – that we are prepared to continually expose our findings, theories and conclusions to doubt. It is also important to acknowledge that, even if we cannot test the correspondence of our propositions with the phenomenon in question, this does not mean that there is no connection between them (Peirce 1878).

Action and experience

Action and experience are key concepts in pragmatism. The terms must be understood in a wider sense than their normal usage. Pragmatism has an aversion to dichotomies, especially the dichotomy that splits body from soul into two distinct spheres, and considers body and consciousness to be inseparable, seeing both the sensory and the idea as part of action. In utilitarianism and in rational choice theories (which might, somewhat simplistically, be considered successors to Descartes' division between body and soul), action is considered from an "evolutionary perspective". In other words, external factors "trigger" the individual's action. This action is based on the individual's pre-established wishes and perceptions of usefulness. From this perspective, therefore, actions can be explained on the basis of (1) external factors and (2) the individual's pre-established ideas, which are considered to be more or less stable. Pragmatism, by contrast, considers the action to be the reaction to a specific sensation in a specific situation (Joas 1993: 18ff). Here, the situation leads us to draw on a part of our reservoir of experiences and expectations from other situations, and to recycle them in order to deal with the current situation. As a result, the meaning of an action is never predictable.

Peirce emphasises that the action's significance lies in its consequences and the other actions to which it leads. We are less interested in looking back and studying the nature of that which created the situation, because the importance of the action is not established until the actual tangible action takes place. Only the consequences of the action are capable of giving us an idea of what the action actually means. The essence of the phenomenon is its consequence. The focus is thus more on the process and possible futures. Similarly, intentionality is not understood teleologically (i.e. as a process that follows a set plan), but as a process that is shaped by the relationships involved – in which both we ourselves, and the actual situation, are constituent parts. Actions are, therefore, always relational.

Today, pragmatism's ideas about action and experience are associated in particular with Peirce's compatriot John Dewey (1859–1952), who was a philosopher, psychologist, educational theorist and practitioner. In particular, Dewey played a prominent role in educational research. He took Peirce's and James's ideas about action and experience a step further. Central to both is their insistence that experience and abduction are the fundamental starting points for learning and knowing about something that is not already known. All sensations arise as reactions to *something* that is only explicable through culturally and historically determined discourses, which in turn must always be based on a physically based sensation of *something*. In this way, the experience is at one and the same time both individual and social (Dewey 2008:

42). Learning is therefore created individually, socially and on the basis of the actual situation. We always learn from the experience itself *and* the social situation in which it takes place or our previous experiences. Where Peirce's interest in abduction concerns any forms of cognition, Dewey's abduction focuses on ways of achieving the aims of education – i.e. the way in which new knowledge is acquired.

Dewey considers education as a means to achieve intellectual freedom, which must not be confused with bodily freedom or the right to do whatever you feel the urge to do (Dewey 2008: 64). For him, freedom does not consist of following your impulses in a given situation, but of being able to master and assess the consequences of your actions, and thereby acting in the best possible way in relation to the initial impulse (Dewey 2008: 67). Mastery therefore implies that, based on an experience, we identify the situation in which we find ourselves and understand its impact on this basis. The only way we can know the meaning of a situation is as a result of past experiences and habits. As such, if we find ourselves in a familiar situation, we do not have to stop and think about how our actions will affect the situation. We are able to rely on the force of habit. Unknown situations, on the other hand, bring cognitive operation into play, and require us to use and reflect on past experiences' similarities with the current situation. We must observe the context in which the experience takes place, and compare it to our knowledge of what we have previously experienced in similar situations and contexts. We then make a qualified assessment of the relationship between the observation of context, similar past experiences and other knowledge. In other words, we conduct an abductive assessment of the situation.

The individual and the social

Most of the American psychologist and philosopher George Herbert Mead's (1861–1931) writings were collected by his students after his death, and facilitated the further development of his thinking. Mead's writings laid the foundation for the approach known as *symbolic interactionism*, formulated by Herbert Blumer in 1937. Symbolic interaction understands behaviour as the process in which the participants, through interaction, negotiate and suggest ways of understanding social situations. Mead's central importance lies in his refinement of pragmatics – in particular, his formulation of the relationship between the individual and the social, and consequently the relationship between habits and creativity.

Mead (1967) divides the self into two parts: an "I" and a "me". In Mead's words, the me is the social self, while the I is a response to the me. The me is the reflexive self, which relates to the social situation, the expectations we think others have of us, and our role in the situation. The me is a cognitive and reflexive object only known retrospectively, which reflects on the social situation and the expectations of the role of the individual concerned, as well as on how the I fulfils these roles. The I's actions are a direct reaction to what is taking place, and will always be a bit different than the last time the individual concerned was in the same situation. The I is therefore creative and free – it acts and creates change.

Identity and the self emerge from relational links to others, and both the I and the me play a crucial role in this. From Mead's perspective, the I is the immediate response to other people's actions, while the me is an organised set of social roles and codes that we incorporate (Mead 1967). Although Mead talks about the me and the I on the individual level, his theory does not focus solely on this aspect. The relationship between the immediate I and the me who focuses on expectations ("the generalised other") takes place in a continuous negotiation, which Mead calls *transactions*. These transactions involve the ability to continuously change the relationship and actions. They are not, therefore, defined in advance by the social situation, nor are they based solely on the individual. Mead formulates the relationship between the individual and the social as a consistent and indivisible whole understood via *both* habitual actions that reflect the social context in which they occur, *and* actions that move and change the social framework by virtue of continuous reinterpretation in reaction to the specific situation.

Mead's distinction between "I" and "me" establishes a description of the relationship between the individual and the social that, firstly, does not make one of the parties an actor – in other words, either the individual's intentions create the social or the social structures determine the individual's potential for action. Secondly, the distinction helps to explain how people act on the basis of habits and social rules, *while* individuals can also break these rules and habits and create real changes via their actions.

Establishment of the subject field in a pragmatic perspective

The subject field in pragmatism consists of actions and experiences. As described above, pragmatism's emphasis on action is based on a physical and cognitive experience that is simultaneously individual, social and reflective. Actions are not teleological, as they are in utilitarianism – in pragmatism, actions are not based on a pre-established objective by which we attempt to achieve the best possible outcome in relation to the situation we are in. Nor are actions, as in structuralism, defined or controlled by mechanisms and structures, and therefore removed from the individual's experiences and ideas about the future. Pragmatism looks at actions quite differently.

In pragmatism, the action always takes place in a particular situation, in which the individual draws on a wide range of experiences, on symbolic meanings, social roles and imperatives from previous situations, in order to deal with the current situation and the consequences of the new act. To understand a specific action's meaning, it is necessary to understand its future consequences. In this context, meaning is established by the action taken by the individual.

For example, the pragmatist Horace M. Kallen (1882–1974), in his article "Democracy versus the melting-pot" (1915), argues that the immigration to America of many different nationalities would not lead to a multicultural "melting pot", and that its significance would only be understood by looking at the consequences of immigration. Kallen therefore studies the consequences of immigrants' actions. In doing so, he demonstrates that, rather than gather in class-segregated groups, thereby strengthening a class-based society, or forming part of a single nation, people congregated according to national orientation. For example, all of the Irish immigrants, both workers and managers, attend the same churches. But these are not the same churches attended by Italian immigrants, even though both groups are Catholics. Kallen's thinking was later credited with showing that mass immigration to the USA, instead of resulting in a multicultural "melting pot", led to a culturally pluralist society, better described as "the salad bowl". In the United States today, there is still a tendency to congregate in churches based on nationality. For example, the Danish Seamen's churches bring together Danes and other Scandinavian immigrants, and act as much more than a religious institution.

One of pragmatism's distinctive characteristics is its anthropology. Pragmatism considers the individual's intentions to be procedural, relational and situational, and therefore both individual and social at the same time. Thus, pragmatism does away with the distinction between the idealistic and materialistic. People are active participants in the social world, based on tangible practices in which previous experiences from other contexts are used to cope with new concrete situations. People always act relationally to other people. Actions are transactions between social roles and more independent identities. Social expectations are used in specific situations, and by specific individuals, who not only embody but also transcend some of these expectations. Actions therefore always involve both something individual/creative and something social/habitual. In pragmatism, the knowledge-constitutive interest is usually ideographic. In order to encapsulate transactions and the relationships between individuals and their social context, analyses are often conducted at the micro level.

Pragmatism cannot be characterised as exclusively realistic or constructivist. There is a constructivist element, because the focus is on the consequences of actions in practice, but there is also a strongly realistic element, because the conditions are assumed to exist objectively in the world. The world is created continuously on the basis of specific events to which we are required to relate. However, that which is created also exists out there and can be studied.

The processual perspective of phenomena is absolutely central to pragmatism. Knowledge of phenomena's significance and meaning is thus not determined once and for all, but is located processually in relation to the specific context of which it is a part. However, the specific situation is not open to all possible interpretations. We might say that there is always "a something" that is known – even if this knowledge changes depending on the experiences the individual concerned brings to the situation. This means that the understanding of the situation is not predefined through previous experiences, but is established in the encounter with that which is constituted by the situation. As a result of this, the pragmatic study is not directed at habits and breaches of habits but at the forms of social meaning and dynamics that arise in the process. Pragmatism, therefore, has an interest in studying the social processes that influence the emergence of social action, and how social action is expressed in various relationships and forms of negotiation.

Example of a research question

In our case study of poverty and the World Bank, we could study the strategy for change adopted in the light of the growing criticism from the 1980s through to the early 1990s triggered by events such as the disaster in *Sardar Sarovar (Narmada) Dam* in India in 1985. All stakeholders, including donors, recipients, NGOs and administrators, agreed that the strategy had to change and address the criticisms that had been raised from various quarters – both generally, regarding the war on poverty, and, more specifically, about project management. However, the World Bank's new strategy, rolled out from the late 1980s to the mid-1990s, encountered obstacles and resistance, culminating in 1996 when James D. Wolfensohn took over the management of an organisation in crisis without a coherent strategy reinforced by its staff in practice.

A pragmatic perspective would be interested in finding out how the different practices in the World Bank interact with or counteract each other, the effect of this on the organisation's social meaning and practices, and the impact on the World Bank's aid strategy. A possible question might be:

- How did the World Bank's social actions and donation practices affect its strategy from the late 1980s until the mid-1990s?

A pragmatic approach would pay particular attention to all of the unexpected consequences of specific practices and actions that have stymied the implementation of the proposed strategy, or at least those that have not resulted in a coherent organisation with a clear identity and strategy designed to unite all stakeholders. The focus would be on how the dynamic between differing social views has led to new actions and new processes. As such, we might be interested in uncovering the social dynamics between the various departments within the World Bank and between the different stakeholders. As the actions studied would be seen as very broad, but at the same time social and individual, it would be normal to study the different actors' social actions, positions in the organisation and social identities, as well as the organisation's practices and structure, in order to map out negotiations about social meaning in the organisation. Likewise, the study would identify how the dynamics between the different social understandings helped to drive certain social practices forward and obstruct others – for example, how the group of social scientists brought in by McNamara from 1977 onwards, and the members of the environmental department created in 1987, established social relationships inside the organisation and negotiated the distribution principles and practices.

Knowledge creation in a pragmatic perspective

The pragmatic perspective assumes that all knowledge stems from the act of sensing something – we do not access the world directly but via signs that represent this something (the world) in different ways. As the meanings of actions are only identifiable through the consequences of those actions, it is important that we adopt a forward-looking perspective when considering phenomena. A pragmatic scientist will therefore investigate the meaning of changes by looking at the specific situation in which they took place and their consequences. This means that social practices provide an obvious vantage point from which to study specific actions and their consequences. On the basis of the specific action, we cannot identify the experiences that must be involved in order to give the action meaning, nor can we imagine the potential consequences of the action. We must also, therefore, map the negotiations of meaning and expectations within the situation itself. In this way, the pragmatic approach is somewhat archaeological or historical in nature, in that it seeks to map the various logics, trends and relationships that characterise a given context. In order to identify the dynamics between repetition and innovation (habits and breaking habits), it is often necessary to work on a micro level, where these opposing movements can be observed and mapped.

As we touched on earlier, pragmatism's analytical practice utilises *abduction* in order to identify an as yet unknown object. Here, small, even minute traces of the situation and context that we want to study are observed and used to hazard guesses and postulate hypotheses that must subsequently be proved. Abduction is about daring to say something that is not immediately apparent, but which seems likely in the given situation if we draw on our knowledge of issues and situations that resemble the phenomenon we are trying to describe. It is therefore important that, in the initial phase, the analysis is open to all possible kinds of explanations and understandings. The second phase consists of the selection and systematic testing of probable explanations. The assessment of the results and conclusions in the pragmatic perspective are closely linked with the ability to explain and understand the context. In this perspective, therefore, knowledge is not value-free. We are not interested in the ultimate truth, but in the most effective way of saying something true about the situation in which we find ourselves.

Example of data acquisition and processing

Considering the question of how the World Bank's social actions and donation practices affected its strategy process from the late 1980s to the mid-1990s, we could focus on the real-life actions and negotiations that took place, and how they were subsequently discussed – i.e. their later significance within the organisation. One approach would be to look at how the strategy not only perpetuates previous habits and frameworks of understanding, but also creates new approaches and practices. This would include studying which transactions between habits and new perspectives are more or less important in the strategy process, and how they contribute to the ongoing construction of meaning.

Such an approach could involve both observation and interviews. Observation could be used to study interpersonal transactions, identify behaviour patterns in the different groups and describe the dynamics that maintain both habits and differences. Specifically, it would be useful to observe collaborative practices associated with the processing of donations. In order to identify these, it is essential to first describe the World Bank's procedures for donations. It would also be natural to look at the specific negotiating practices in relation to which the World Bank's strategy is discussed and formulated, as well as the general practices associated with allocating aid. By being made on a number of levels, the observation serves to identify how the different practices lead to contradictions and co-operation, and how this collaboration specifically occurs in the tension between what the individual or group does (the I), and the way the individual/group reflects society's (partners') expectations (the me).

Interviews could, for example, be used to reveal how the individuals/groups relate to expectations from the social context, and therefore how individuals/groups' past experiences are employed when dealing with specific situations. This would contribute to the mapping of the many – often contradictory – approaches to the same situation, and the different experiences involved in achieving an understanding of them. Interviews could also be used to identify the similarities and differences in symbols used by the individuals/the groups to describe and explain the various transactions and relationships.

To be able to study the impact of individual actions going forward, it is necessary either to conduct a study over a longer time-scale or to include previously produced material (typically in writing) that can provide insight into how similar situations were handled in the past. We could, for example, look at the reasons behind the incorporation of the two new groups in the World Bank (social scientists and environmental managers) into the process of defining the practices and strategies that form the basis for the aid principles. Policies, minutes of meetings and recorded objections could also be used as data material.

The pragmatic analysis, based on an abductive approach, uses the material collected to identify different possible understandings – i.e. "qualified guesses" – about the social processes that take place, and then engage in a more systematic study of which guess is the most probable or, after close examination, needs to be rejected. Here, the empirical material will be used at various stages: firstly, to make the abductive guess; then to describe the internal dynamics of the situation being studied; and finally, to assess whether the analysis holds water. The analysis builds on its explanatory power and usability. In other words, only once every stone has been turned does it make sense to define the relationship between the World Bank's strategy process and donation practice in a specific way, and determine whether it is a rewarding way to analyse and describe the organisation and whether it adds new knowledge/understanding of the phenomenon. If so, then from a pragmatic perspective the analysis is usable, and therefore true, until its truth is challenged.

CHAPTER 10

Actor-network theory and the return of materiality

In recent decades, actor-network theory (ANT), cultural geography and a closer focus on materiality have made greater and greater inroads into a host of disciplines, offering new and different angles from which to raise and approach research questions. A key feature of this development has been the detailed studies of how certain ideas and phenomena make an impact, spread and attain the status of truth in specific contexts. These studies aim to find a form of grammar that describes the way in which different parts of the world are interrelated.

ANT has an element of constructivism, in that it looks at how certain things become truth(s). However, there is also an element of realism in ANT, as it considers not only social structures, but also phenomena and purely physical and biological conditions. A well-known example is Bruno Latour and Steve Woolgar's *Laboratory Life* (1979), which examines how scientific facts emerge – namely, through a combination of biochemical and biological findings in the laboratory and the academic rules regarding the publication of scientific articles, in which scientific facts are propagated until they acquire the status of truths. In practice, ANT closely analyses specific contexts, and is materialistic in the sense that the point of interest is in why different things, people, ideas, values, rules, etc. look the way they do in specific contexts.

It is important to emphasise that the ANT perspective is not a school, and that it has been applied in a range of highly diverse ways in very different areas, from scallop fishing and the invention of the vaccine to economic conditions and the great Portuguese naval explorers. ANT is still developing, and doing so rapidly, making it a very new perspective compared to others described in this book.

ANT traces its roots back to a range of other theories, including semiotics, the French Annales School, symbolic interactionism, Gilles Deleuze and Félix Guattari's rhizome thinking and Michel Foucault's concept of power. The perspective gained ground in science and technology following Latour and Woolgar's laboratory studies. It spread in the 1980s following pioneering empirical analyses by, among others, Bruno Latour, Michel Callon, and John Law. In the 1990s, ANT broadened out from studies of technology to other areas – particularly, organisational studies. In the 2000s, it expanded into research fields such as medicine, information and communications technology, education, learning and economics. It now plays an increasingly important role as a source of inspiration in most social science and humanities disciplines.

ANT should not be thought of as a complex of theories, but as a particular methodological approach that is sensitive to empirical studies, and which has consequences for the questions we raise and the answers we receive. The best ANT analyses extend over a prolonged period, provide a detailed description and mapping of how obvious facts (referred to as black boxes) arise, and list the various networks (social, biological, economic, etc.) associated with these black boxes. The challenge is to identify how obvious facts derive their status as truisms.

Overall, we can generalise about ANT as follows:

1. ANT challenges *the unity of science* in the positivist sense but adds a new concept of unity, in which all areas and facts can be studied through their emergence via networks and associations. This applies to all kinds of phenomena, irrespective of whether they are defined as part of the social, human or natural sciences.
2. The subject matter therefore comprises these heterogeneous networks and their emergence. Networks can be stud-

ied and described but do not form the basis for transposing generalisations to other networks, as each network has its own genesis. ANT analyses are therefore ideographic.

3. The ontological basis for ANT is that the world is as it is – both in a realistic and constructivist sense. This means that the more we construct, the more real that which we construct becomes. In other words, the wider and larger the network that creates the black boxes, the more real the constructed facts (black boxes), because the networks become increasingly distant from their original heterogeneity.

4. ANT's epistemology is associated with description and mapping. It uses an ethno-methodological approach, i.e. an anthropologically inspired methodology in which dynamics and interaction are considered in a specific context. The contexts and explanations for phenomena are created through analysis of how the actors involved weave their way in and out of the various networks and make connections between different interests. In other words, ANT follows the traces left by the actors in the network, and tries to unravel how their tracks interweave across various spheres.

5. Although ANT aims to map specific and objective conditions, its basis is still that science can never be *value-free*.

6. The view of human nature that characterises ANT is based on actants. The actants performing the action are not necessarily humans but are also things and other non-human actors. Humans never act alone but always in interaction with other (including non-human) actors. The view of humankind is therefore relational and anti-essential, so the focus is on actions, not intentions.

7. The truth theory that applies in ANT is based on the pragmatic concept of truth. In other words, something is considered to be true when the results are useful and helpful to explain and understand phenomena and events. Our perception is true as long as it can be used to understand

the specific situation being studied and it is not necessary to be able to generalise the points (as is required in logical positivism).

Starting point and sources of inspiration for ANT

ANT grew out of science and technology studies (STS), which draws on a tradition from Karl Marx to Robert K. Merton's (1910–2003) sociology of science, as well as Thomas Kuhn's history/philosophy of science and the Edinburgh School. In their own way, each dealt with the relationship between science and the social world and raised questions about the role the social world plays in relation to the development of scientific disciplines, findings and truths. In recent years, the STS tradition has grown exponentially as a separate discipline and is to be found in different versions based on a variety of traditions. Its focus on the relationship between the social world and scientific approaches continues to gain ground in still more disciplines.

Merton's sociology of science, developed in the early 1960s, did not include ideology and relativism as subject areas. Rather, he turned away from the more cognitive aspects of science to focus on the social aspects. In turn, he divided the sociology of science into external and internal science conditions, which meant that scientific discoveries and knowledge could be considered as epistemologically different from the social and human sciences, and therefore the natural sciences were not explained in social terms.

Kuhn's work in the 1960s, by contrast, raised awareness that the historical and social elements were important for both the form of the sciences and their cognitive core – in other words, the evaluation of which scientific conclusions can be considered true, and why (see Chapter 3).

In the mid-1970s, the Edinburgh School – David Bloor, Barry Barnes, Steven Shapin and others – extrapolated radical epistemological consequences from Kuhn's work. They formulated the "strong programme", in which external and internal science conditions and what are considered true and false statements must be studied and explained in the light of social factors. Bloor calls this the symmetry principle. The strong programme insists that no disciplines or researchers – not even in the natural sciences – can be placed outside the social construction of knowledge. The Edinburgh School therefore equates the natural sciences with the social and human sciences, and insists that their form and content have to be studied and explained in the same way.

The idea of studying the construction of knowledge in all sciences, as promoted by ANT and STS, therefore has a long history. At the same time, there is no doubt that the broad field of actor-network theory, cultural geography and materialisation theories must also be seen as a reaction against, and a response to, the criticisms and shortcomings of social constructivism's triumphal progress through the academic disciplines. A common criticism of social constructivism is that it has a unilateral focus on linguistic constructions and interpretations that resolve all discussions of *why* by referring to the constitutive role of language and "the social". Why are immigrants oppressed? Because the discourse about immigrants creates a negative inclusion of immigrants in the social space.

Social constructivism's companion was political emancipation, and this led to great productivity in deconstructing, historicising and culturalising of all sorts of social and cultural phenomena, ultimately resulting in the "open work" (Eco 1989) that became social constructivism's epistemological dictum. "Work" here should be interpreted broadly as "meaning" – the point being that meaning does not exist in itself, but is continually created through social and linguistic interactions, which in turn means that all significance and meaning can be changed and made to take a fairer direction. This may sound nice, but scientifically it is not optimal, because certain things in the world do have quite specific meanings and opinions, regardless of people's attitudes – most will agree, for example, that the meaning of "molecules" or "the diesel engine" is not really open to interpretation.

Another criticism of social constructivism is that, as a result of its focus on liberation, it is now the preferred instrument of power and consumption, a point highlighted in Latour's "Why Have Critiques Run Out of Steam?" (2004). According to Latour, the tragedy of social constructivism is that it has created its own dark side through its relativist critique of scientific realism's universality. This has led to the abolition of the reality concept, only to be replaced by a discussion of how realities vary across different spaces. The assumptions underpinning this relativism are found in a split between the various (almost) closed groups, each of which is able to insist on its own reality and perspective – because, of course, the truth is dependent on which camp you are in. However, according to Latour, this mindset blurs and limits real insight into and understanding of how and why the world looks the way it does.

Key concepts in ANT

Spokespeople

Rather than being either realism or social constructivism, ANT is a "new realism", as Latour (1993) calls it. This is a realism that does not divide phenomena according to the Western academic world's preferred dichotomies: culture/nature, language/thing, micro/macro, subject/object. According to Latour, such dichotomies reflect a dualism, which means that we have already lost sight of how phenomena in the world really are – in the real world, nothing is 100% one thing or the other. The dualism is touched on in Latour's *Science in Action* (1987), but is then properly considered in *We Have Never Been Modern* (1993). The latter not only shows how dualism has been conjured up as the epitome of modernity and of the Western understanding of science since Descartes but also how, through continuous scientific divisions into dualistic concepts, it has produced even greater hybridity rather than abolishing it.

Although, on the face of it, many dichotomies appear to be

part of the scientific purification process, they actually create additional hybridity and instability. The best-known example cited by Latour to illustrate the lack of division between the natural and the social is precisely the scientific purification process that has taken place since the Enlightenment. Consequently, the more we have insisted on dividing the world into clear and fixed dichotomies, the more heterogeneous and messy the phenomena found in those dichotomies have become. The abolition of sociological dichotomies is therefore central to ANT, which takes as its starting point that reality is both real and constructed, subjective and objective, specific and universal, etc. The more created reality becomes, the more real it is too.

The abolition of dichotomies reflects a rethinking of science, which also involves a critique of the theory fetishism that plays a key role in the social sciences and, through social constructivism, in cultural analyses. The consequence of theory-driven analysis has mainly focused on describing and explaining events and actions through meta-theoretical approaches that seek to identify "underlying frameworks" and supporting logic, rather than on the actions and events in all their complexity. As a result, contradictory and unexpected emotions, impulses and fates are excluded from explanations and descriptions because they are irrelevant to the theory concerned. Similarly, much of modern social science uses the context and its logic to explain and justify events, which are therefore considered as effects of the context. The material turn and, in particular, ANT are characterised by following how things/events are generated when different networks are linked together by different human and non-human actors across areas, levels and cognition points (Latour 1987). Therefore, the empirical analyses and descriptions are central to the material turn. The analyses express the uniqueness of the perspectives. In the following paragraphs, we will look at a number of researchers and studies that have helped to define ANT, all of which work with more or less the same concepts and precepts.

Translation

One important exponent of ANT is Professor of Sociology Michel Callon, whose main area of interest is technological- and natural-science projects and the relationship between economic techniques, economists and the economy. As a result, Callon's approach is particularly suited to the analysis of transactions, in a broad sense. His interest in transactions forms the basis for his thinking, and he has described ANT as a translation sociology that defines a new approach to power, in which science and technology play a key role in structuring power relationships (Callon 1986). "Translation" is a central concept in ANT. It is used by all ANT-inspired researchers and originates from the French philosopher Michel Serres (1977). Translation occurs when phenomena move from the sphere in which they originated, in which they take a particular form and are understood, to another sphere in which they will be endowed with a different meaning and play a different role. The translation processes will often lead to a simplification of the phenomena's heterogeneous conditions and in doing so negotiate and limit the identities of the actors involved, forms of interaction and options for acting in new ways.

In Callon's ground-breaking article on scallop fishing in Northern France, "Some elements of a sociology of translation: Domestication of the scallops and the fishermen of St. Brieuc Bay" (1986), he shows how a study focusing on where and how translation takes place is able to uncover the uncertainties and ambiguities that are produced and balanced out through translation processes. In the article, Callon formulates the ANT perspective's critique of the idea that the social should be able to explain all – and especially all scientific – phenomena. In response to social constructivism, Callon describes three methodological approaches, which form the basis for ANT thinking:

1. The first he calls agnosticism, which implies that the identity of the phenomena in question is not known in advance. It is created through the network and translations that we

are interested in analysing, and therefore it is important not to fix it in advance.

2. Consequently, Callon reformulates the Edinburgh School's symmetry principle – that is, the idea that all of science's internal and external factors must be studied on the basis of the same principles. Where the Edinburgh School's "ordinary" STS approach uses only social explanatory factors, Callon extends the symmetry principle to a "generalised symmetry principle", which implies that the same conceptual vocabulary is retained in descriptions of all of the phenomena being studied (see as well Latour 2005: 76). When we study contexts, this means that we should not divide them up into social and material spheres, but consider all of the elements involved in the construction of facts and phenomena. This is unlike the approach taken by the Edinburgh School, in which the social aspect is generally considered the only one that is useful for explaining or describing developments.

3. Through the use of what he calls "free associations", Callon advocates avoiding drawing distinctions between the social and the natural because the two are separated by means of translation processes, and accordingly cannot be used as an explanatory framework. The difference between the natural and the social is produced tangibly (in Callon's article, through the network between scallops, fishermen and scientists). If we are interested in the dynamics that characterise scallop fishing, it is inappropriate to take a predetermined contrast between the social and the natural as the starting point. It is more beneficial to look at how this contrast is produced and how that process impacts on the way in which fishing works.

In the article, Callon describes how three types of human and non-human actors are linked together in a network through translation processes that stabilise their conflicting interests: 1) the fishermen, who would like to increase the stock of scallops to ensure a revenue basis; 2) the biologists, who want to understand the scallops' reproductive and generative processes; and 3) the scallops, which are endangered. In this light, Callon formulates four central processes that most ANT researchers use to a certain degree in their description of the translation processes that build up tangible heterogeneous networks:

- First comes the "problematisation", in which the problem is defined as a common starting point to which all of the actors involved can and must relate in order to promote their interests. This makes the actants who formulate the starting point indispensable actors, and their problematisation serves as an "obligatory passage point". It is simply impossible to participate in the context without acknowledging an interest in the joint starting position.
- Next comes "interessment". This involves the actants trying to recruit different actors to their project, as a result of which the various actors are "locked into" specific roles and identities defined by the project. In Callon's article, the fishermen are locked into maintaining the basis for their revenue, researchers are locked into finding out about scallop reproduction patterns, and the scallops are locked into trying to survive and flourish in Saint-Brieuc Bay.
- Then an "enrolment" takes place, during which the different roles and identities are defined and associated with each other and accepted by the different actants. Other "counter-programmes" can challenge this enrolment, in an attempt to organise other networks and offer other obligatory passage points. In this case, the scallops are enrolled by the researchers' attempt to get them to establish themselves from larvae in the bay until they grow large enough to be consumed. This process is challenging. In this specific example, the researchers found they could not get enough scallops to settle on the banks established for them.
- Once a "winner" has emerged from this struggle between the different programmes, "mobilisation" can take place.

Mobilisation involves the negotiation, identification and subsequent mobilisation into action of spokespersons who represent and simplify the interests of the different passive collective groups. The negotiation involves establishing common definitions and opinions regarding the phenomena that link the network, the issues discussed, who speaks on whose behalf and whether that representation is fair and acceptable. A successful negotiation and translation establishes a strong actor-network with non-competing interests. In this case, it is the scallop larvae captured by biologists, and their behaviour, which becomes the "spokesperson" for all of the scallops. The fishermen have a voting procedure to choose their spokesperson. The research community is represented by the three biologists who have defined the problem pertaining to the scallops in the bay.

The spokesperson, which may or may not be human (e.g. it might be a map recording marine life or a biologist who talks about the scallops' reproduction process at a conference), simplifies the interests and actors that form the network, which enables the actors' interests to be moved to different spheres than the one from which they emerged. However, there is a constant risk that the individual spokesperson does not speak on behalf of the masses and cannot mobilise them, or is challenged as spokesperson and replaced by another. Therefore, any network is always unstable. Any translation involves change and displacement. The creation of networks and the establishment of spokespersons make it possible to simplify and move facts, knowledge, ideas, concepts and understandings from the context in which they were created to other contexts, where they enter into new networks.

Actors and positions

Professor of Sociology John Law prefers to describe ANT as "material semiotics". As such, he stresses ANT's semiotic and relational characteristics, in which the elements of a network are constantly defined and create each other through "translation". ANT is, in Latour's words, "a semiotic definition of how wholes are built" (Latour 1999b: 7f), i.e. the methodological perspective is concerned with how instability and heterogeneity, which are the basic characteristics of the world, are fixed and included in building worlds.

One example of this can be seen in Law's article, "On the methods of long-distance control: Vessels, navigation and the Portuguese route to India" (1986). In it, he examines how Portuguese domination and the connection between centre and periphery was practised for 150 years during the 1500s and 1600s, through the simplification and mobilisation of three kinds of groups of apparently unstable actors: (1) documents about navigation, geography and astronomy; (2) regulations on, e.g. shipbuilding (here, the Carreiras), and navigation instruments such as the quadrant (the forerunner of the sextant) and the astrolabe; and (3) trained people, including sailors, astronomers, geographers and merchants. These networks translated and displaced knowledge, interests and identities via spokespersons and inscription apparatuses.

Inscription apparatuses are a central concept in ANT. They describe rules that simplify and pin down unstable and heterogeneous networks and compress them into a single expression or a single rule that can subsequently be transported to spheres other than those from which it originated, and that can be linked to new actors (Jensen, Lauritsen & Olesen 2007). In this case, an inscription could be the quadrant or astrolabe, devices that – without comprehensive understanding of the knowledge behind their production – can be used for navigation in relation to latitude and longitude, based on the constellations and the position of the sun. In Callon's example, it could be a graph that simplifies and presents figuratively large amounts of data regarding the scallops' reproduction patterns. Behind both inscriptions are large and long networks consisting of heterogeneous actors who through the inscription apparatuses can be easily mobilised and transferred to new contexts.

Central to Law's description of how the Portuguese mastered the oceans is the use of complicated devices that have a simple and practical importance in a completely different context. Ship technology led to large, reliable, fast and safe vessels with small crews. New navigation techniques, which brought together practical and astronomical knowledge of the use of latitude and longitude, and included manuals and devices, made it possible to navigate without land in sight. Law's article describes how, through translations, the different units made unstable and heterogeneous networks into a unit that extended and united their own interests with those of others across spheres, and in doing so established dominance. His study is about how power is actually exercised and how it involves a heterogeneous mass of actors and demarcations, which together create a structure characterised by the undisturbed communication and mobile, persistent, strong actors essential for long-distance control.

Semiotics is a major inspiration in Law's work. As he writes: "Actor-network theory is a ruthless application of semiotics" (Law 1999: 3). Semiotics is absolutely central to ANT's definition and understanding of actors. The starting point here is the French semiotician Algirdas Julien Greimas' actantial model. In his analysis of narratives across time and space, Greimas (1970) established a number of main positions/functions (recipient-object-giver-helper-subject-opponent) that recur and form a basic narrative structure. Greimas began by analysing fairy tales, in which there is usually a hero who is sent out on a mission, meets various opponents and, at journey's end, receives a reward. He generalised the model to a language theory that applies not only to fairy tales, but to understanding narratives and narration in general (in the linguistic sense, but also in all other sign systems). The key point here is that it is the relation to other positions that defines each position's meaning and substance and establishes identities and action – i.e. the hero is a hero because he is a contrast to a villain.

Therefore, the positions do not have, a priori, a substance or an identity that defines their relation to other identities. They derive their identity by entering into and acting as a network with other actors. Actants are defined by their function and can be both concrete and abstract (as in Law's example of the quadrant or Portuguese dominance), and human and/or non-human actors (both documents and sailors). An actor can easily assume multiple positions in the network. Actants are therefore established in the actor-networks, and this generates both challenges and opportunities. In his first ethnographic laboratory studies, Latour used an actant analysis to describe the practices that led to the laboratory's truth statements (Latour & Woolgar 1986; Latour 1993). The actor and actant concepts are absolutely central to ANT. The concepts of actant and actor are often used interchangeably to emphasise that non-human actants act just as much as human actors do, even if intentions cannot be attributed to them (Latour 1996: 373). As Latour puts it: "An 'actor' in ANT is a semiotic definition – an actant – i.e. something or somebody that acts or has actions associated with it by another actor" (Latour 1999b: 24). The relational understanding of the actant forms the basis for studying all actants, human and non-human, in the same way as they potentially contribute to shaping the network and closing what Latour calls "black boxes".

Black boxes and actor networks

Since the millennium, Bruno Latour has achieved almost superstar status in the academic world. His studies encompass philosophy and anthropology, and he has placed ANT at the centre of thinking about science and the nature of science in all fields. His name invariably crops up whenever ANT is discussed. Latour wrote several of the pioneering case studies that characterise ANT. In a multitude of texts, he describes the thinking behind ANT, with reference to both case studies and more theoretical texts.

As mentioned earlier, Latour is interested in the translation mechanisms that create actor-networks and bring together heterogeneous actants and multiple interests in "black boxes" – that is, closed truisms that appear clear and homogeneous. In Latour's terminology, "black boxes" consist of any combination of ideas, things and people that adds input and receives output without questioning how the output was actually generated. One example is Latour's study of the diesel engine, in which he shows how ideas about new ways to create propulsion, e.g. via ignition and the struggle of Rudolf Diesel and his colleagues to develop and spread their concept, are brought together to become the diesel engine, which is then enclosed in a black box that appears to be a matter of fact: diesel is pumped in as input; out of the other end comes the propulsion of cars, locomotives, etc. As Latour describes it:

> When a machine runs efficiently, when a matter of fact is settled, one need only focus on its inputs and outputs and not on its internal complexity. Thus, paradoxically, the more science and technology succeed, the more obscure and opaque they become. (Latour 1999a: 2104)

The establishment of actor-networks has a temporal character. Most of Latour's studies (and the majority of ANT-inspired studies) focus on opening existing black boxes or study how black boxes are established and help to keep unstable and heterogeneous networks together. In his landmark book *The Pasteurization of France* (1984), Latour describes the formation and closing of the black box around Louis Pasteur's bacteria research, which led to the invention of the vaccine that protected livestock from deadly diseases and, later, to the development of techniques for pasteurisation and combating bacteria. Latour shows how these facts are the consequence of the establishment of an actor-network that led to Pasteur's discoveries spreading as far and wide as they did. In his work with microbes, Pasteur moved his laboratory out to a small French village, where an anthrax outbreak had killed large numbers of animals. He collected microbes and bacteria, took them home to his laboratory in Paris and developed a vaccine. Because the vaccine worked, he went back to the village, gave it to the sick animals and saved much of the herds. He then returned to Paris, where he enrolled health professionals in the fight against bacteria, aiding them in their struggle for better hygiene among the poor. This whole process was followed with great interest by politicians and journalists, so that the movement between the laboratory and the world was continually reported and therefore served to translate, displace and move interests and actors between different spaces. This meant that the storm-tossed and often very unclear creation myth of bacteria was represented as a homogeneous system of clear units and truths.

Latour demonstrates how Pasteur managed to capture the interests of others – the farmers, the microbes, the animals and the health professionals – and enrol them into his research. He did so through repeated displacements and translations that shifted the interests of others so that they were united in a common goal – counteracting bacteria. By dint of this enrolment, Pasteur established his research into bacteria as the obligatory passage point that brought together apparently conflicting interests and was able to formulate and meet all of the actors' interests. At the same time, it established him as the spokesperson for the microbes, researchers, politicians and farmers. In the process, Pasteur closed the black box that made the bacteria the main cause of various disasters in France, which therefore had to be fought in every conceivable way, on every conceivable level.

Latour's ethnographic description is a textbook example of how an ANT perspective leads us to study the way in which specific heterogeneous actor-networks are strategic, relational and productive over time. Central to the filling up and closing of black boxes is the fact that the movement between what goes on inside (the laboratory) and outside (politics, farming, dead animals) forms the basis for translating and transforming the various actors' identities. As such, Latour emphasises that we cannot divide our studies into micro- and macro-actors because all actors get their identity *via* the networks. In other words, the laboratory, agriculture, the microbes, Pasteur himself, the sick animals, etc., end up looking and acting in a certain way because of the actor-network in which they take part. The starting point must therefore be that there is no difference between subject and object. Rather, someone or something differentiates subject and object. The mediator – in this case, Pasteur – is the figure that produces the difference.

As mentioned earlier, Latour's use of ethnographic methods forms the background to the main features of the specific form of analysis in ANT studies. In his book *Science in Action* (1987), he formulates five basic principles, including the mantras "Follow scientists in action" and "Follow the history of the statement/fact". These absolutely central methodological approaches involve a close description of the processes that take place during the creation and maintenance of unstable networks, which enable us to describe how they transcend the division between inside/outside or micro/macro, and at the same time help define how this distinction is defined. His third basic principle is that we cannot determine the actors' identity in advance, but must observe the manner in which it is determined. Latour's fourth and fifth methodological basic principles are that neither the "social" nor the "natural" should be considered the main ingredient in an explanation of controversies and processes, but as constructions in the actor-network.

Establishing a target field in an ANT analysis

As discussed earlier, Latour and ANT critique both positivist ideals and social constructivism, based on an ontological position that the facts are true and constructed at the same time, and that the more constructed they are, the more real they become. The argument is that facts are produced as black boxes through translations involved in the formation of actor-networks. The extent of the actor-network has significance for the strength (and hence reality) of the network and the facts. Long networks may contain multiple black boxes and enrol more interested parties, which are brought together via spokespersons who establish a few obligatory passage points, in doing so ensuring greater consistency between the heterogeneous actors in the network and their different interests. This criticism of social constructivism and positivism is based, in particular, on the fact that they allow external factors or additional context to serve as explanation for the phenomena being described. From an ANT-perspective, neither the social nor the natural can be used as an explanatory factor. By focusing on the process by which phenomena and facts are constructed, ANT also criticises the positivist approach's understanding that science describes how things are in reality. ANT is interested in describing reality, but argues that it cannot be described as a set of rules of general application, in the way that positivism wishes (at least in its most consistent form).

By focusing on materiality, ANT clashes with the idealism that has characterised much of constructivism – particularly social constructivism. ANT is interested in material, heterogeneous relationships that have left visible imprints. These therefore have to be observed, rather than derived from analyses. ANT is interested in how unstable networks are held together in practice, and how these networks help to give each component its specific form and shape, and also produce specific practices. Actor-networks are generally analysed through case studies and the application of ethnological and semiotic tools. They can only be evaluated and described ideographically, rather than generalised across different case studies. This is partly due to ANT's take on contexts. In ANT, unlike hermeneutics, contexts are not used to explain phenomena. Rather, the contexts emerge from the analysis – in other words, they are established through the study and not described in advance as a premise or explanatory framework. The point is to describe what is actually happening, which means accounting for the relevant and specific context via the analysis – it cannot be assumed in advance.

ANT is interested in actants – which are not only people, but can also be non-human actors. By concentrating on actants, ANT focuses on everything that has created and creates action: Scallops cling to the banks, the quadrant shows the way, the ships sail, the microbes move, and Pasteur studies, talks, writes and reaches agreements. There is a difference between intentional and non-intentional actants, but this is not important because the intentions are only studied in relation to the tracks left by the actors. Unlike hermeneutics and pragmatics, an ANT analysis is not interested in mapping and describing people's experiences and expectations, but only in identifying the traces they have left.

Example of a research question

In our case study of the World Bank, it would be possible to study the bank's funding principles from an ANT perspective – for example, via these two focus points:

- How the multiple voices that exist in the organisation are organised through translations and enrolments when looking at the funding principles.
- How the boundary between the organisation and the outside world is brought into play and constantly challenged when looking at the funding principles.

In the first example, we are looking at closing a black box. There are different groups with different interests in the World Bank. The appointment of McNamara and the consequent building up of a group of social scientists in 1977 represents one group with specific interests and understandings of problem-solving. Another is Clausen's environmental group, established in 1987. An ANT study would examine how these groups' interests have been displaced and brought together around a common goal, a common "problematisation" regarding the relationship between combating poverty and socio-economic development.

When we examine the period 1970–2000, we see a progression in how this problematisation is defined. An ANT analysis would therefore look at how the problematisation is determined when specific groups bring together the different (basically heterogeneous) interests in an actor-network, in which they make themselves indispensable in the fight against poverty and in doing so construct an obligatory passage point. The idea is to study how the groups concerned recruit not only other groups, but also the techniques used to calculate and evaluate the extent of the aid needed and its impact. This "locks" the various actors into specific roles that are defined by the project, and an "enrolment" takes place. The study would focus on how the different roles in this translation and displacement are defined and associated with each other and accepted by the different actants, so that their place and their identities in the network are defined and can be mobilised into action via their spokespersons.

This displacement and translation often meet fierce resistance – and sometimes the network will collapse. In our case study, we see that the unstable actor-network collapses several times, because it is too short and not strong enough to cope with the displacement of interests that takes place – in other words, the black box is too unstable to keep closed. An example of this arises when McNamara is replaced by Aden Clausen, and new actants (NGOs and media coverage) and new calculation principles have to be translated into the existing network. On several occasions, this leads to a change in the obligatory passage point – the explanation of the relationship between social and economic development and the fight against poverty. An ANT-inspired analysis of this type will also describe the requirements that these multiple voices and instability place on the management of the World Bank.

The second question, which investigates how the boundary between the organisation and the outside world is brought into play through the funding principles, considers how the World Bank continuously constructs its environment, how the boundary between outside and inside is displaced and how, in doing so, new interests are constructed. Here, it is relevant to study NGOs and the media. Which new points of view are established when NGOs and the media's definition and identity shifts from outside to inside? Which actor-network is needed to construct new interests via a displacement and translation of actants? How is this significant for the construction of the obligatory passage points and new spokespersons?

Knowledge creation in an ANT-inspired analysis

ANT focuses on conditions that are observable, and therefore on conditions, phenomena and events that have in some way left traces that we are able to follow (Latour 2005). That which does not leave traces is not a part of the final network. This therefore contrasts with the hermeneutical tradition, which provides the basis for much historiography. Interpretation is replaced by an ethnographic description, a description that carefully and closely follows the networks of identifiable trails and connections, which are revealed by this retrospective mapping exercise, the preferred starting point for which is the opening of various black boxes.

A network is connected by the actors, whose agency affects the network's movement and development. ANT's credo is therefore: "Follow the actors in action" and "Follow the history of the statement/fact". The phenomena's context and explanations are created by the actors' involvement in the various networks, as well as how these actors weave in and out of the various networks and bind different interests together. That is why Callon, and ANT researchers in general, emphasise the use of "agnosticism" in the study, i.e. that we do not know about the phenomenon in advance. This means that we must always let the phenomenon's identity and context be defined via the analysed network and translations, instead of establishing them in advance. At the same time, it is important in an ANT analysis to maintain the "generalised symmetry principle", by means of which we describe different phenomena with the same vocabulary in order not to insert or maintain a difference that is actually established via the actor-network. An ANT study does not explain why something is the way it is. It observes how things become what they become and therefore are. The evaluation of the study is based on its ability to explain and describe. The study therefore has a pragmatic truth concept, in which knowledge is true as long as it can be used to understand the specific situation being studied.

Example of data acquisition and processing

In our example of the heterogeneous networks that construct the past and present funding principles for the World Bank's development aid, an ANT study would draw up a very close ethnographic description of the processes involved in the attempts to close the black box. By retrospectively opening and dismantling the actor-network, an ANT approach would clarify the different displacements, interest formations and mobilisations that take place in the translations. Here, the core principles "Follow the actors in action" and "Follow the history of the statement/fact" offer a way to untangle the apparently tangled network. This requires that truisms and identities in the network are considered as a process that is created through the network and facilitates the continuous creation of entities, rather than something that is defined once and for all. This applies both to the struggles that take place between, for example, the various groupings of social scientific staff and to the initiatives taken by the environmental group or the NGOs; and to the media or the political actors used in the donor or recipient countries in an attempt to enrol other actants in the network by offering a useful obligatory passage point.

In our example, therefore, we will be able to follow the attempts to establish and close the black box that asserts that poverty can be combated by economic initiatives. We will also see how attempts are made to open this black box by what might be called counter-programmes, which make other actors spokespersons and establish other obligatory passage points (e.g. other non-economist groups may attempt to foster the mindset that social reforms, rather than aid, should be the starting point for combating poverty and boosting the economy). Instead of assuming in advance that a certain understanding of combating poverty applies in the context concerned, an ANT analysis will look at how certain understandings are cemented via the actor-network, which is capable of creating representations that can be moved from their original sphere to other, larger spheres and possibly back again (as Pasteur did with his movement between the laboratory and farmers' cowsheds), and thus move the world with a lever, as Latour puts it.

The study could, for example, uncover how the boundary between the World Bank and the outside world is displaced and identified in relation to obligatory passage points, e.g. the famine in the South Sahara. Here, we would closely follow the story, the facts and the actors in order to identify which contexts, and therefore explanations, are used to create identities for the actors and weave them in and out of the various networks, connecting different interests together. Since explanations are generated through networks, it will be absolutely crucial to maintain the generalised symmetry principle and look at how explanations translate separate concepts and divide up the understanding of development aid into categories, as well as how these concepts and categorisations are included as actants in the network. In this context, all actants – both human (such as the various groupings in the World Bank) and non-human (such as logistical models for calculating aid or policy documents and concepts) – must be described by a common vocabulary. The study will only be able to uncover how the unstable network is held together in a particular case, and therefore it will be an ideographic study that cannot be generalised to other organisations, as their translations, mobilisations and actants would be completely different.

References

Archer, M.S. (1995). *Realist Social Theory: The Morphogenetic Approach.* Cambridge: Cambridge University Press.

Archer, M.S. (1998). *Critical Realism: Essential Readings.* London: Routledge.

Archer, M.S. & J.Q. Tritter (2000). *Rational Choice Theory: Resisting Colonization.* London: Routledge.

Berger, P.L. & T. Luckmann (1966). *The Social Construction of Reality.* New York: Anchor Books.

Bhaskar, R. (1978). *A Realist Theory of Science.* Leeds: Leeds Books.

Bhaskar, R. (1989). *The Possibility of Naturalism: A Philosophical Critique of the Contemporary Human Sciences.* Brighton: Harvester Press.

Bhaskar, R. (1998). Societies. In: Archer, M. (ed.) *Critical Realism: Essential Readings.* London: Routledge.

Bloor, D. (1976). *Knowledge and Social Imagery.* London: Routledge & Kegan Paul.

Blumer, H. (1969). *Symbolic Interactionism: Perspective and Method.* Englewood Cliffs: Prentice-Hall.

Callon, M. (1986). Some elements of a sociology of translation: Domestication of the scallops and the fishermen of St. Brieuc Bay. In: Law, J. (ed.) *Power, Action and Belief: A New Sociology of Knowledge?*, pp. 196–233. London: Routledge & Kegan Paul.

Carnap, R. (1934). *Logische Syntax der Sprache.* Wien: Julius Springer.

Christensen, G. (2002). *Psykologiens Videnskabsteori: Introduktion.* Frederiksberg: Roskilde Universitetsforlag.

Coddington, A. (1972). Positive Economics. *Canadian Journal of Economics*, 5(1): 1–15.

Comte, A. (1830-1842). *The Course on Positive Philosophy*, vol. 1–6. London: J. Chapman.

Cunningham, D.J., J.B. Schreiber & C.M. Moss (2005). Belief, Doubt and Reason: C.S. Peirce on Education. *Educational Philosophy and Theory*, 37(2): 177–189.

David, M. (1994). *Correspondence and Disquotation: An Essay on the Nature of Truth*. Oxford: Oxford University Press.

Delanty, G. (2005). *Social Science: Philosophical and Methodological Foundations* (2. ed.). Maidenhead: Open University Press.

Delanty G. & P. Strydom (2003). *Philosophies of Social Science: The Classic and Contemporary Readings*. Buckingham: Open University Press.

Descartes, R. (1641). *Meditations on First Philosophy*. South Bend: Infomotions.

Derrida, J. (1976/1967). *Of Grammatology*. Baltimore: Johns Hopkins University Press.

Dewey, J. (2008/1916). *Democracy and Education*. Radford: Wilder Publications.

Dilthey, W. (1883). *Einleitung in die Geisteswissenschaften*. Leipzig: Duncker & Humblot.

Dilthey W. (1989). *Selected Works, vol. 1: Introduction to the Human Sciences*. Princeton: Princeton University Press.

DiMaggio, P.J. & W. Powell (1991). *The New Institutionalism in Organizational Analysis*. Chicago: University of Chicago Press.

Durkheim, E. (1982/1895). *The Rules of Sociological Method*. New York: The Free Press.

Eco, U. (1989). *The Open Work*. Cambridge: Harvard University Press.

Feyerabend, P.K. (1975). *Against Method: Outline of an Anarchistic Theory of Knowledge*. London: NLB.

Foucault, M. (1995/1975). *Discipline and Punish: The Birth of the Prison*. New York: Random House.

Foucault, M. (2009/1961). *History of Madness*. New York: Routledge.

Foucault, M. (2002/1966). *The Order of Things*. New York: Routledge.

Friedman, M. (1979). The Methodology of Positive Economics. In: Hahn, F. & M. Hollis (ed.) *Philosophy and Economic Theory*. Oxford: Oxford University Press.

Gadamer, H.G. (1986–1990). *Wahrheit und Methode. Grundzüge einer philosophischen Hermeneutik.* Tübingen: J.C.B. Mohr.

Gergen, K. (1994). Realities and Relationships. Soundings in Social Constructionism. Cambridge: Harvard University Press.

Glanzberg, M. (2009). Semantics and Truth Relative to a World. *Synthese,* 166(2): 281–307.

Glanzberg, M.J. (2013). Truth. In: Zalta, E.N. (ed.) *Stanford Encyclopedia of Philosophy.* Stanford: Metaphysics Research Lab.

Greimas, A.J. (1987/1970). *On Meaning.* Minneapolis: University of Minnesota Press.

Greimas, A.J. & J. Courtés (1982/1979). *Semiotics and Language: An Analytical Dictionary.* Bloomington: Indiana University Press.

Guba, E. (1990). *The Paradigm Dialogue.* Beverly Hills: Sage.

Guba, E. & Y.S. Lincoln (1994). Competing Paradigms in Qualitative Research. In: Lincoln, Y.S. & N.K. Denzin (ed.) *Handbook of Qualitative Research.* London: Sage.

Habermas, J. (1987/1968a). *Erkenntniss und Interesse.* Frankfurt am Main: Suhrkamp.

Habermas, J. (1968b). Technik und Wissenschaft als "Ideologie". *Merkur,* 22(243): 682–693.

Hacking, I. (2000). *The Social Construction of What?* Cambridge: Harvard University Press.

Harré, R. (1986). Varieties of Realism: A Rationale for the Natural Sciences. Oxford: Basil Blackwell.

Heidegger, M. (2007/1927). *Being and Time.* Aarhus: Klim.

Hobbes, T. (1651). *Leviathan.* Oxford: Clarendon Press.

Hume, D. (1739–1740). *A Treatise of Human Nature.* Oxford: Clarendon Press.

Husserl, E. (1936). *Die Krisis der Europäischen Wissenschaften und die Transzendentale Phänomenologien.* Amsterdam: Martinus Nijhoff.

Jackson, F. (2006). Representation, Truth and Realism. *The Monist,* 89(1): 50–62.

James, W. (1907). *Pragmatism: A New Name for Some Old Ways of Thinking.* New York: Longman Green & Co.

Jantzen, C. (1996). *Selvfølgeligheder: Kulturanalyse som Praksis: Om Kulturanalyse som en Tekstanalytisk tilgang til Enkeltfænomener, som Ana-*

lytikeren er Samtidig Med. Aalborg: Institut for Kommunikation, Aalborg Universitet.

Jenkins, K. (1991). *Re-Thinking History*. London: Routledge.

Jensen, B.E. (1986). The Role of Intellectual History in Dilthey's "Kritik der historischen Vernunft". *Dilthey-Jahrbuch für Philosophie und Geschichte der Geisteswissenschaften*, 2.

Jensen C.B., P. Lauritsen & F. Olesen (2007). *Introduction to STS: Science, Technology, Society*. København: Hans Reitzels Forlag.

Joas, H. (1993). *Pragmatism and Social Theory*. Chicago: University of Chicago Press.

Kallen, H.M. (1915). Democracy versus the Melting-Pot: A Study of American Nationality. *The Nation*, 100 (2590): 190–194.

Kant, I. (2002/1781). *Kritik af den Rene Fornuft (Critique of Pure Reason)*. Frederiksberg: Det lille Forlag.

Kuhn, T.S. (1962). *The Structure of Scientific Revolutions*. Chicago: University of Chicago Press.

Kuhn, T.S. (1970a). Normal Science as Puzzle-Solving. In: Kuhn, T.S. & I. Hacking (ed.) *The Structure of Scientific Revolutions*, pp. 35–45. Chicago: The University of Chicago Press.

Kuhn, T.S. (1970b). The Nature and Necessity of Scientific Revolutions. In: Kuhn, T.S. & I. Hacking (ed.) *The Structure of Scientific Revolutions*, pp. 92–111. Chicago: The University of Chicago Press.

Laclau, E. (1996a). Why Do Empty Signifiers Matter to Politics? In: Laclau, E. (ed.) *Emancipation(s)*, pp. 36–46. London: Verso.

Laclau, E. (1996b). The Time is Out of Joint. In: Laclau, E. (ed.) *Emancipation(s)*, pp. 66–83. London: Verso.

Laclau, E. & C. Mouffe (1985). *Hegemony and Socialist Strategy: Towards a Radical Democratic Politics*. London: Verso.

Lakatos, I. (1970). History of Science and its Rational Reconstructions. *Boston Studies in Philosophy of Science*, 8: 91–135.

Lakoff, G. & M. Johnson (1980). *Metaphors we Live By*. Chicago: University of Chicago Press.

Latour, B. (1984). *The Pasteurization of France*. Cambridge: Harvard University Press.

Latour, B. (1987). *Science in Action: How to Follow Scientists and Engineers through Society*. Cambridge: Harvard University Press.

Latour, B. (1993). *We Have Never Been Modern*. Cambridge: Harvard University Press.

Latour, B. (1996). On actor-network theory: A few clarifications. *Soziale Welt*, 47(4): 369–381.

Latour, B. (1999a). Give Me a Laboratory and I Will Raise the World. In: Biagioli, M. (ed.) *The Science Studies Reader*, pp. 258–276. London: Routledge.

Latour, B (1999b). On recalling ANT. In: Law, J. & J. Hassard (ed.) *Actor Network Theory and After*, pp. 15–25. Oxford: Wiley-Blackwell.

Latour, B. (2005). *Reassembling the Social: An Introduction to Actor-Network-Theory*. Oxford: Oxford University Press.

Latour, B. & S. Woolgar (1986/1979). *Laboratory Life: The Social Construction of Scientific Facts*. London/Beverly Hills: Sage.

Law, J. (1986). On the methods of long-distance control: Vessels, navigation and the Portuguese route to India. In: Law, J. (ed.) *Power, Action and Belief: A New Sociology of Knowledge?* London: Routledge and Kegan Paul.

Law, J. (1999). After ANT: Complexity, naming and topology. In: Hassard, J. (ed.) *Actor-Network Theory and After*, pp. 1–14. Oxford: Blackwell.

Lawson, T. (1997). *Economics and Reality*. London: Routledge.

Lévi-Strauss, C. (1967). *The Savage Mind*. Chicago: University of Chicago Press.

Lewin, K. (1951). *Field Theory in Social Science: Selected Theoretical Papers*. D. Cartwright (ed.). New York: Harper & Row.

Locke, J. (1689). *An Essay Concerning Human Understanding*. Oxford: Clarendon Press.

Luhmann, N. (1993). Deconstruction as Second-Order Observing. *New Literary History*, 24(4): 763–782.

Lynch, M.P. (2009). *Truth as One and Many*. Oxford: Oxford University Press.

Marx, K. (1857–1858). *Grundrisse der Kritik der Politischen Ökonomie*. Frankfurt: Europäische Verlagsanstalt.

Marx, K. (1867, 1885, 1892). *Das Kapital*. Hamburg: Meissner.

Marx, K. & F. Engels (1848). *Das Kommunistische Manifest*. Berlin: Buchhandlung Vorwärts.

Mead, G.H. (1967/1934). *Mind, Self and Society*. Chicago: University of Chicago Press

Merleau-Ponty, M. (2005/1945). *Phenomenology of Perception*. London: Routledge.

Merton, R.K. & N.W. Storer (1973). *The Sociology of Science: Theoretical and Empirical Investigations*. Chicago: The University of Chicago Press.

Mill, J.S. (1843). *System of Logic: Ratiocinative and Inductive: Being a Connected View of the Principles of Evidence, and the Methods of Scientific Investigation*. London: J.W. Parker.

Moore, G.E. (1953). *Some Main Problems of Philosophy*. London: George Allen & Unwin.

Peirce, C.S. (1877). The Fixation of Belief. *Popular Science Monthly*, 12: 1–15.

Peirce, C.S. (1878). How to Make Our Ideas Clear. *Popular Science Monthly*, 12: 286–302.

Peirce, C.S. (1931). *Collected Papers of Charles Sanders Peirce, vol 1*. Cambridge: Harvard University Press.

Peirce, C.S. (1932). *Collected Papers of Charles Sanders Peirce, vol 2*. Cambridge: Harvard University Press.

Peirce, C.S. (1992). The first Rule of Logic. In: Ketner, K.L. (ed.) *Reasoning and the Logic of Things: The Cambridge Conferences Lectures of 1898*. Cambridge: Harvard University Press.

Popper, K. (1959/1935). *The Logic of Scientific Discovery*. London: Routledge.

Popper, K. (1972). *Objective Knowledge: An Evolutionary Approach*. Oxford: Clarendon Press.

Rorty, R. (2007). *Philosophy as Cultural Politics: Philosophical Papers, vol. 4*. Cambridge: Cambridge University Press.

Russell, B. (1910). On the nature of truth and falsehood. In: Russell, B. (ed.) *Philosophical Essays*, pp. 147–159. London: George Allen & Unwin.

Saussure, F. de (1974/1916). *Course in General Linguistics*. London: Fontana.

Sayer, A. (1984). *Method in Social Science: A Realist Approach*. New York: Routledge.

Sayer, A. (2000). *Realism and Social Science*. London: Sage.

Serres, M. (2000/1977). *The Birth of Physics*. Manchester: Clinamen.

Serres, M. & B. Latour (1995). *Conversations on Science, Culture and Time*. Ann Arbor: University of Michigan Press.

Shook, J.R. & J. Margolis (2005). *Companion to Pragmatism*. Malden: Blackwell.

Smith, B.H. (2005). *Scandalous Knowledge: Science, Truth and the Human*. Durham: Duke University Press.

Snow, C.P. (2001/1959). *The Two Cultures in Science*. Cambridge: Cambridge University Press.

Walker, R.C.S. (1989). *The Coherence Theory of Truth*. London: Routledge.

Weber, M. (1995/1904). *Den Protestantiske Etik og Kapitalismens ånd (The Protestant Ethic and the Spirit of Capitalism)*. Copenhagen: Nansensgade Antikvatiat.

Young, J.O. (2001). A Defense of the Coherence Theory of Truth. *Journal of Philosophical Research*, 26(1): 89–101.

Zahavi, D. (2003). Fænomenologi. In: Collin, F. & S. Køppe (ed.) *Humanistisk Videnskabsteori*. Søborg: DR Multimedie.

Case references

Cernea, M. (2004). *Culture? At the World Bank? Letter to a Friend*. Find it at: cultureandpublicaction.org/pdf/cernealet.pdf.

Cochrane, S. & R. Noronha (1973). *A Report with Recommendations on the Use of Anthropology in Project Operations of the World Bank Group*. Washington: World Bank.

Cornia, G.A., R. Jolly & F. Stewart (1987). *Adjustment with a Human Face, vol. 1–2*. Oxford: Oxford University Press.

Danaher, Kevin (1994). *50 Years is enough: The Case against the World Bank and the International Monetary Fund*. Boston: South End Press.

Davis, G. (2004). *A History of the Social Development Network in the World Bank, 1973–2002*. Washington: World Bank.

Fox, J.A. (1998). When Does Reform Policy Influence Practice? Lessons from the Bankwide Resettlement Review. In: Fox, J.A. & L.D. Brown (ed.) *The Struggle For Accountability: The World Bank, NGOs and Grassroots Movements*, pp. 303–344. Cambridge: MIT Press.

Kapur, D., J.P. Lewis & R. Webb (1997). *The World Bank: Its First Half Century*. Washington: Brookings Institution.

Kardam, N. (1993). Development Approaches and the Role of Policy Advocacy: The Case of the World Bank. *World Development*, 2(11): 1773–1786.

Lateef, S.K. (1995). The First Half Century: An Overview. In: Lateef, S.K. (ed.) *The Evolving Role of the World Bank: Helping Meet the Challenge of Development*, pp. 1–36. Washington: World Bank.

MIGA (1998). *Annual Report*. Washington: MIGA.

Miller-Adams, M. (1999). *The World Bank: New Agendas in a Changing World*. London/New York: Routledge.

Morse Commission (1992). *Sardar Sarovar: The Report of the Independent Review*. Ottawa: Resources Futures International.

OED (1995). *Learning from Narmada*. Washington: World Bank.

Rich, B. (2002). The World Bank under James Wolfensohn. In: Pincus, J.R. & J.A. Winters (ed.) *Reinventing the World Bank*, pp. 26–53. Ithaca: Cornell University Press.

Wapenhans, W.A. (ed.) (1992). *Effective Implementation: Key to Development Impact, Portfolio Management Task Force*. Washington: World Bank.

World Bank (1994). *Resettlement and Development Report*. Washington: World Bank.

World Bank (1995). *Wolfensohn Lays out Future Direction of World Bank*. Washington: World Bank.

World Bank (1998). *Global Development Finance, Analysis and Summary Tables*. Washington: World Bank.

Glossary

Anomalies
Phenomena that cannot be explained by a specific paradigm or theory. Too many anomalies can lead to the collapse of the current normal science.

A posteriori
Knowledge occurs *after* the experience/sensing of the world.

A priori
Knowledge occurs rationally and *before* the experience/sensing of the world.

Causality
Phenomena have direct causes, referred to as causation.

Coherence theory
The degree of truth in a proposition is based on the relation to other propositions in a coherent system.

Contextual view
A contextual way of looking at phenomena is the belief that we must take into account the temporal, spatial and cultural contexts in which they are embedded.

Correspondence theory
Something is true when there is consistency between scientific propositions and empirical reality. A statement's validity is tested by studying whether it is consistent with the real world.

Cumulative knowledge
The idea that our knowledge is built on existing knowledge, and we therefore become wiser and wiser.

Deduction
The movement from theory to result, and thus a theoretical framework for arranging observations. Conclusions are drawn from general rules to specific cases.

Demarcation criteria
The distinction between science and non-science.

Epistemology
The study of cognition and knowledge. Epistemology is about the nature of knowledge, how we can know, and how knowledge can and should be produced.

Epoché
The central analytical concept in phenomenology. The researcher places parentheses around theoretical knowledge and doubts all preconceptions, in order to observe how phenomena appear to the subjects.

Falsification
A concept in critical rationalism in which empirical studies are used to test and contradict own hypotheses.

Functionalism
This concept is typically explained on the basis of an organism model, in which each individual part can only be understood in terms of the role it plays in order for the organism to exist as it does.

Ideographic
The particular or peculiar, where each event or action is considered in its own specific context and has its own specific history and genealogy.

Incommensurability
This concept refers to different paradigms being incomparable, and therefore not cumulative.

Individual reduction
An understanding of society and other units as consisting of an aggregation of individuals.

Induction
Proceeding from the empirical case to universal laws and relationships. General rules are based on conclusions drawn from individual cases.

Intersubjectivity
Multiple researchers are able to reach (approximately) the same result using the same analyses and interpretations.

Methodological reduction
The idea that all subject areas must be studied in the same way in order to be scientifically described, even if they are not ontologically identical.

Nomothetic
The analysis of phenomena in order to establish general laws or regularities about the nature of the phenomenon.

Normative
How something *should* be, rather than how it is.

Objectivity (methodological)
Methodical separation of the person who would like to gain

knowledge and what they want to know about. The subject (the researcher) looks at an object (what he or she wants to know about, which can be separated from the self and considered in isolation).

Ontological reduction

The idea that all objects generally have the same kind of being – they are essentially the same – and can be studied in the same way.

Ontology

The study of being. Our basic assumptions about the nature of the world and the things that exist in it.

Paradigm

A paradigm is both a world view, i.e. a general shared idea of what the world is and can be, and a view of science based on universally accepted scientific propositions, which, at least for a while, indicates exemplary research problems and solutions.

Pragmatic concept of truth

Something is true when it is useful or fruitful, either practically or scientifically. In other words, when something is useable, it is true (until proven otherwise). Thus we are dealing with a relationship between people, institutions and propositions.

Processual view

A way of looking at phenomena that recognises that they are dynamic and changeable. In other words, we can never describe reality fully, as phenomena are continuously evolving.

Subjectivity (methodological)

Methodical use of the relation between what you want to gain knowledge of and the thing you want to know something about. One subject (the interpretive researcher) looks at another subject (what you want to know something about, and which cannot be separated from the interpreter).

Unity of science
The idea that all science is based on the same principles, in terms of either ontology or methodology.

Universal view
A universal way of looking at phenomena is the belief that phenomena do not change across time and place. This is closely related to the positivist perspective.

Value freedom – three positions
(1) We must be objective/value-neutral, and the study should be kept separate from the researcher and the context.
(2) It is not possible to separate the researcher from the study and be completely objective.
(3) The research should take into account politics and values, and therefore be useable/have a purpose.

Verification
The search for confirmation of identically established hypotheses/theses via empirical observations.

View of human nature
The study of humankind. What are people, and what drives them?

Index

Matrix: Presentation of perspectives in the philosophy of science

	Logical positivism	Critical rationalism	Hermeneutics	Phenomeno-logy	Structuralism
Purpose The know-ledge-constitutive interest.	Study of causal relationships and predictions based on them.	Study of causal relationships and predictions based on them.	Study of hu-man-generated opinions and meanings.	Study of hu-man conscious-ness and cogni-tion, including the way phe-nomena appear to humans.	Study of how unconscious and invisible structures form the basis for collective meanings and actions.
Ontology On what basic assumptions about the (social) nature of the world is the individual perspective based?	Realistic onto-logy. Phenome-na have univer-sal essence.	Realistic onto-logy. Phenome-na have univer-sal essence.	Realistic ontology. Phe-nomena must be understood contextually.	Essence is existence. Abolition of the dualism be-tween realism and construc-tivism. Pheno-mena must be understood contextually.	Realistic onto-logy (mainly). The phenome-non is studied for its univer-sal essence.
Views of humanity and agency	Humans are considered to be utilitarian and rational. Studies are based on indi-vidual reduc-tion (except the Durkheim tradition).	Humans are considered to be utilitarian and rational. Studies are based on indi-vidual reduc-tion.	Humans are intentional beings. The importance and meanings the individual attributes to phenomena and events are considered to be contextual.	The individual is an inten-tional being who directs attention and actions towards something, with intention behind every action.	Human beings are subject to structural conditions. The individual's intentions are not individual, but part of the collective.

	Logical positivism	Critical rationalism	Hermeneutics	Phenomenology	Structuralism
Epistemology What is the nature of knowledge, how can we know, and how can – and should – knowledge be produced?	Knowledge is achieved a posteriori through an inductive approach and must be verified. The purpose is to explain as nomothetically and objectively as possible.	Knowledge is achieved a priori, via a deductive approach, and has to be falsified. The purpose is to explain as nomothetically and objectively as possible.	Knowledge is about understanding and can be achieved both deductively and inductively. The purpose is to understand ideographically, via subjective interpretations.	Knowledge is about understanding and is achieved a posteriori, via an inductive approach. The purpose is to understand ideographically through intersubjective interpretation.	Knowledge is about understanding and explaining nomothetically, and can be achieved both deductively and inductively via intersubjective analyses.
Concept of truth How do we assess the validity of propositions?	Value freedom is the ideal. Truth is evaluated by correspondence. A relationship between the proposition and the world.	Value freedom is the ideal. Truth is evaluated by correspondence. A relationship between the proposition and the world.	Value freedom is impossible and undesirable. Truth is evaluated by means of coherence theory, whereby propositions are considered true if they can be included in a system of interpretive statements and not be contradicted.	Value freedom is impossible and undesirable. Truth is evaluated by means of coherence theory, whereby propositions are considered true if they can be included in a system of interpretive statements and not be contradicted.	Truth is evaluated through a kind of correspondence theory, where propositions are considered true if it is probable that the structures uncovered actually exist independently of the researcher's interpretation.

	Critical realism	Social constructivism	Discource analysis	Pragmatism	Actor-network theory
Purpose The knowledge-constitutive interest	Study of how unconscious and invisible mechanisms and structures form the basis for collective meanings and actions.	Study of how ways of thinking and speaking, as well as everyday truisms, are established, used and changed.	Study of how discourses establish object and subject positions.	Study of how actions and experiences from previous situations affect and are used in the current actions, and the potential consequences of this.	Study of how different translations are created and take place in heterogeneous networks.
Ontology On what basic assumptions about the (social) nature of the world is the individual perspective based?	Realistic ontology. The phenomenon is studied for its processual essence.	Constructivist ontology. Phenomena do not have essence per se – we study how they emerge processually.	Constructivist ontology. Phenomena do not have essence per se – we study how they emerge processually.	Neither realistic nor constructivist ontology. Phenomena are studied processually and their meaning is determined by their impact.	Both realistic and constructivist ontology. This means that the more we construct, the more real that which we construct becomes.
Views of humanity and agency	Humankind is subject to structural conditions and mechanisms. The individual's intentions are not individual, but part of the social and can be changed through the continuous use of social mechanisms.	The individual is considered as a representative of a general collective, defined by discourse, social constructions and structures.	The individual is considered as a representative of a general collective, defined by discourse, social constructions and structures.	People's intentions are considered processual, relational and situational, and thus both individual and social.	The view of humankind is relational and anti-essentialist. The starting point is actants, both human and non-human. Therefore, the focus is on actions, not intentions.

	Critical realism	Social con-structivism	Discource analysis	Pragmatism	Actor-network theory
Epistemology What is the nature of knowledge, how can we know, and how can – and should – knowledge be produced?	Knowledge is about understanding and explaining nomothetically, and can be achieved through retroduction.	Knowledge is about understanding inductively how meaning is ascribed to phenomena and describing how this process works. The purpose is to understand ideographically, via subjective interpretations.	Knowledge is about understanding inductively how meaning is ascribed to phenomena and describing how this process works. The purpose is to understand ideographically, via subjective interpretations.	Knowledge is about understanding and explaining the meaning of phenomena ideographically, and is achieved a posteriori through abductive analyses.	Knowledge is achieved by inductively following actants and history. Description and mapping is done via an ethno-methodological approach and ideographically.
Concept of truth How do we assess the validity of propositions?	Truth is evaluated through a kind of correspondence theory, where propositions are considered true if it is probable that the structures uncovered actually exist independently of the researcher's interpretation.	Truth is evaluated by means of coherence theory, whereby a proposition is true if it can be included in a system of interpretive statements and cannot be contradicted.	The individual is considered as a representative of a general collective, defined by discourse, social constructions and structures.	Truth is evaluated through the pragmatic concept of truth, which is a *relationship* between people, institutions and propositions. Something is true when the results of a study help to explain phenomena and events.	Truth is evaluated via the pragmatic concept of truth, which is a *relationship* between people, institutions and propositions. Something is true when the results of a study help to explain phenomena and events.